Construction
Math's Explained

Robert Cooke

Builda books UK

This edition first published 2018

© 2018 Builda Books UK

First published in Great Britain in 2018

The right of Robert Cooke to be identified as author of this work has been asserted by him in accordance with the Copyright, Design and Patents Act 1988.

ISBN: 978-1-9993109-1-2

British Library Cataloguing in Publication Data

A catalogue record for this book is available from th British Library

Printed and bound in Great Britain

About the author

Robert Cooke is a former College lecturer with over 25 years experience of teaching BTEC (EdEXCEL) National and Higher National Construction students.

He has a trade background as a shopfitting carpenter and joiner, site fixer and setter out.

A change of direction to working on the professional side of construction has provided a wealth of practical experience.

A BSc with the Open University and teaching qualifications has helped with the understanding of returning to higher education as a mature student.

Good luck with your studies. We should never stop learning.

Why yet another book about Math's?

Math's – some people find it easy and others, well you're probably one of the others and this book is written for you.

It is for Construction students who are wanting to enter Higher Education. Employers want qualified staff and many clients require it as part of their contract.

You may be very well experienced but don't have the paper qualification so further education or higher education is required; Construction Higher Certificate / Diploma or Degree level.

If it's a few years since you have studied, this book will help improve your math's study skills. Unless you use math's, calculators and computer spreadsheets regularly it doesn't take long to forget how to use them.

The whole concept of this book is to make things straightforward.

Yes that's the idea, a book written in plain everyday language explaining ways of solving problems. Math's should be used as a tool to do a job.

It's not a 'metallurgic arc rotating impact implement' but a hammer.

"Oh dear! The kinetic energy from my arc rotational impact implement has just impacted my thumb".

Many students have said "Why don't I write a Math's book?" – Well here it is.

Table of Contents

Chapter 2 – Rules of math's

Chapter 3 – Profit margin

Chapter 4 – Algebra *"It's a lot of mumbo jumbo to me"*

Chapter 5 – Areas

Chapter 6 – Volumes

Chapter 7 – Irregular volumes

Chapter 8 Right angled triangles

Chapter 9 Alternative trigonometry

1.00 *Numbers*

We haven't always had them and over the centuries we have adopted various systems and added or changed them. They are a code that enables ideas and information to be passed to other people.

Without numbers a heap of nuts and bolts on the floor could be a large heap or a small heap, but no reference to how many there are. Using a code of numbers provides a quantity. The quantity would be written using numbers termed *'whole numbers'*.

There are nine numbers also known as 'units'; 1, 2, 3, 4, 5, 6, 7, 8 and 9. Zero means there are no numbers. When more than 9 units are needed; 9 + 1 for example it is written as 10. Meaning that there is 1 set of 'tens' and no units so a zero is shown.

Yes we all know that so why tell us?

One of the most common causes of mistakes made in calculations is due to missing zeros. People leave off the zeros and when adding, subtracting, dividing or multiplying the columns and get mixed up.

Keep all the units in the column on the right, then tens and so on. The same applies to calculations such as subtraction, multiplication and division.

It also makes life easier for anyone checking the figures.

It is very easy to make a mistake if numbers are written as in column 'A'.

A	B
12	12
127	127
2043	2043
3	3
7513	2185
incorrect	**correct**

Quantities - Example 01

So we now have 527 bolts opposed to a heap or pile.

Whole numbers: 5 sets of 100, plus 2 sets of tens and 7 set of units.

Example 02

Counting the nuts there are 507 in total. Again whole numbers as there are no parts of a nut. There are 5 sets of 100, zero set of tens and 7 set of units. It's important to enter the zero for the tens column otherwise it would appear as 57 which obviously is wrong.

Example 03

There's also a box of washers totalling 530. Whole numbers; 5 sets of 100, 3 sets of tens and zero set of units.

That's 'whole numbers' but what happens when there is a part of a number? It can be written as either a fraction or a decimal of a number.

1.01 Decimals

When writing any whole number there is a theoretical 'point' or dot after the units. Termed the *'decimal point'* it can be written either half way up a number: 2·0 indicating there are 2 units and no parts termed 'decimals'. Or as commonly across Europe, spreadsheets and calculators the decimal point appears as a comma or full stop at the base of the numbers: 2.0

Decimals are grouped in blocks of 10 behind the decimal point ('behind' refers to the right side of the point and 'in-front' of is to the left of the decimal point).

Each column behind the decimal point is termed a decimal place. The first is tenths, then hundredths and then thousandths and so on. Mirrored to the columns of whole numbers.

Fig 1.01 shows the relationship between whole numbers and decimals of a number.

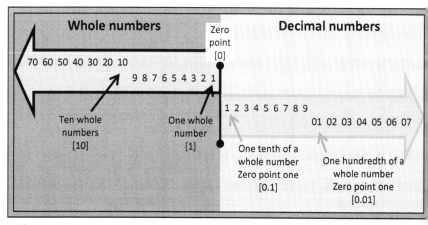

Fig 1.01

In construction lineal measurements are normally in metres. The decimal point indicates the decimals of a metre:

0.1m = 100mm

0.01m = 10mm

0.001m = 1mm.

Note the zero before the decimal shows there are no whole numbers. Other reasons for the zero are:

1. The decimal point may not be noticed if say hand written on a survey sheet. 0 215 would still suggest the measurement is in metres even if the decimal point is not obvious.

2. When entering data on a spreadsheet. Having a zero before the decimal point will keep the columns in line. Spreadsheet formatting will normally put the zero in automatically.

Another method of writing parts of a number is fractions.

1.02 Fractions

Where a whole number is divided up into a number of same sized parts it is fractionalised. Names have been given to the number of parts;

'Half' means there are two equal sized parts.

'Third' is three equal parts

'Quarter' is four equal parts

And so the list goes on. To save writing a fraction in words the number of equal parts (termed the *denominator*) is written below a horizontal line.

How many parts (denominators) are shown above the line termed the *numerator*.

Another term for this type of fraction is *vulgar fraction* or *common fraction*.

1.03 *Adding fractions*

Example 01

If you and two friends order a pizza that has been sliced up into eight equal sized pieces, each would be an eighth. Also be written as $\frac{1}{8}$

The denominator (the number under the line: 8) indicates there are 8 equal sized pieces. Note they have to be eight <u>equal</u> sized to be 'eighths'.

If one friend has two slices ($\frac{2}{8}$), the other friend has three slices ($\frac{3}{8}$), and you have three slices ($\frac{3}{8}$) there will be no pizza left.

To add fractions of the same size just add the numerators (number above the line).

In this case 2 + 3 + 3 = 8. The total of numerators is the same as the denominator: $\frac{8}{8} = 1$.

So adding the numerators of the same sized fractions is straightforward addition. If the total of numerators is bigger than the denominator (an improper fraction) then divide the denominator into the numerator.

$\frac{11}{8}$ is an improper fraction. Divide 8 into 11 goes 1 with 3 left over; $1\frac{3}{8}$

Example 02

Still with pizzas; there is a group of 5 friends all wanting pizza. Knowing each pizza will be cut into eighths and each friend wanting a different number of slices how many pizzas will be required?

Friend One: 3 slices, friend Two: 3 slices, friend Three: 2 slices, friend Four: 3 slices, and friend Five: 2 slices.

Add the numerators. $\frac{3}{8} + \frac{3}{8} + \frac{2}{8} + \frac{3}{8} + \frac{2}{8} = \frac{13}{8}$

Divide the denominator (number under the line) into the numerator. 8 into 13 = 1 and $\frac{5}{8}$. Therefore 2 pizzas will be required and there will be some left over ($\frac{3}{8}$).

It is common practice to reduce the size of a fraction to the smallest applicable. For example $\frac{2}{8}$ could be shown as a smaller number of slices. In this case a quarter ($\frac{1}{4}$). The numerator can be divided into the denominator (2 divided into 8).

1.04 *How can I add fractions of different sizes?*

Example 03

Tidying your shed you come across two 1ltr cans of paint. Both had been previously used. One had 2/3 left and the other was half full.

Is there too much paint to pour into one can so you can throw the other away?

The first stage is to write the two fractions down – fig 1.02.

Then multiply the numerator of one fraction (top) by the denominator (bottom) of the other fraction (3x1). Now multiply the denominator by the other denominator (3x2).

Next stage is to multiply the other fraction numerator by the denominator (2 x 2). As the denominators have been multiplied together the answer must be the same, in this case 6 (2 x 3) which means both of the fractions are now going to be 6ths of a whole number.

The result will be: 3 sixths and 4 sixths ($\frac{3}{6} + \frac{4}{6}$) = 7 sixths ($\frac{7}{6}$).

As 6 sixths is a whole can of paint that means there is 1 whole tin of paint and 1 sixth left in the other can.

$$\frac{2}{3} + \frac{1}{2}$$

$$\frac{2}{3} \,\nearrow\, \frac{1}{2} = \frac{3}{6}$$

$$\frac{2}{3} \,\searrow\, \frac{1}{2} = \frac{4}{6}$$

$$\frac{3}{6} + \frac{4}{6} = \frac{7}{6}$$

$$\frac{7}{6} = 1\frac{1}{6}$$

$$\frac{2}{3} + \frac{1}{2} = 1\frac{1}{6}$$

Fig 1.02

1.05 *Subtracting fractions of different sizes*

Use the same method of converting the fractions into the same denominator then it is a straightforward subtraction.

Example 04

$$\frac{3}{4} - \frac{5}{8}$$

Convert the first fraction to the same size denominator. That will mean multiplying by 2. (2 x 4 = 8). If the denominator is multiplied by 2 then the numerator must also be multiplied by 2 to keep the fraction the same size.

2 x 4 = 8 and 2 x 3 = 6 therefore $\frac{3}{4} = \frac{6}{8}$

The fractions now have the same denominator therefore it is a straightforward subtraction: $\frac{6}{8} - \frac{5}{8} = \frac{1}{8}$

Example 05

$$\frac{7}{8} - \frac{15}{16}$$

Convert both fractions to the same size denominator. 8 will go into 16 two times therefore multiply the 7 by 2 and the 8 by 2.

$$\frac{14}{16} - \frac{15}{16} = -\frac{1}{16}$$

Example 06

$$-\frac{7}{8} - \frac{15}{16}$$

(Rule: minus, minus a minus = a plus. So although the result is still a minus figure the two numerators are added together).

Convert both fractions to the same size denominator. 8 will go into 16 two times therefore multiply the 7 by 2 and the 8 by 2.

$$-\frac{14}{16} - \frac{15}{16} = -\frac{29}{16} = -1\frac{13}{16}$$

1.06 *Multiplying fractions of different sizes*

In maths 'of' means multiply. If you had 3 boxes 'of' screws it would be 3 x 1box = 3 boxes.

Firstly convert to the fractions so they have the same denominator, and then multiply the numerators.

Example 07

$$\frac{2}{3} \text{ of } \frac{7}{8}$$

Multiply the numerators and divide by the total of the denominators (2 x 7) ÷ (3 x 8) = $\frac{14}{24}$ = $\frac{7}{12}$

1.07 *Dividing different sized fractions*

Example 06

What is $\frac{3}{7}$ divided by $\frac{2}{3}$?

Turn the second fraction upside down (invert) and multiply out.

$(\frac{3}{7})$ x $(\frac{3}{2})$

Numerators; 3 x 3 and denominators; 7 x 2. Becomes $\frac{9}{14}$

1.08 Calculators

There are several different makes of calculator available plus most mobile 'phones have a calculator app. However they do not all work in the same way or have the same buttons.

Throughout this volume and the others in the series a Casio *fx – 85GT PLUS* calculator has been used.

> Replay button includes direction by pressing the edges. →↓←↑

It is a relatively cheap instrument that cannot be programmed therefore is normally accepted for use in exams at all levels even at degree.

Fig 1.03

1.09 *Setting up the calculator*

Turn the calculator on and press shift then mode.

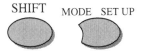

Then press 1and then 1 again to select Mth10.

Repeat using the shift and mode button and 3 to select degrees.

> It is important to ensure the calculator is set up for <u>degrees</u>
> otherwise all the results will be incorrect.

With that set up the fractions button will present one number over another.

The screen results will appear as standard as opposed to decimal. That means the results will be shown as a fraction and not a decimal.

It's easy to change the results to decimal, just press the button (**S**tandard to **D**ecimal).

To change between an angle in degrees, minutes and seconds and decimal degrees, just press the SD button twice. The result will change to degree decimal.

Press the tadpole again and it reverts to degrees, minutes and seconds.

1.10 *Using a calculator*

(Fractions and Decimals)

This section covers how to use the calculator to solve calculations involving whole numbers with fractions and decimals.

It is always good practice to either reset or check the mode the instrument is in. Someone else may have used the calculator and left it in another mode.

Reset: *ON*, then

To check it is in the correct mode press the fraction button (FB) and enter 2.

Then press the down button (↓) It is the bottom part of the large round button – see fig 1.03.

(To save space ↓ indicates the down button and → indicates the right hand button. The fractions button will be FB).

Now enter 3 → +FB 2↓3=

So the whole calculation will be: FB 2↓3→+FB 2↓3= and the answer should appear as $\frac{4}{3}$. That is an improper fraction meaning that it can be converted to a whole number plus a fraction.

Now press *shift* followed by the *SD* button and the answer will convert to a whole number plus the remaining fraction $1\frac{1}{3}$. The *SD* button is 'standard' to 'decimal'.

Check the earlier examples.

Example 01:

$$\frac{2}{8} + \frac{3}{8} + \frac{3}{8} =$$

Calculator: FB 2↓8→+FB 3↓8→+ FB 3↓8=1.

Example 02:

$$\frac{3}{8} + \frac{3}{8} + \frac{2}{8} + \frac{3}{8} + \frac{2}{8} =$$

Calculator: FB 3↓8→+FB 3↓8→+ FB 2↓8→+ FB 3↓8→+ FB 2↓8= $\frac{13}{8}$. Now press *shift*, *SD* and the answer should be $1\frac{5}{8}$.

Example 03:

$$\frac{1}{2} + \frac{2}{3} =$$

Calculator: FB 1↓2→+FB 2↓3= $\frac{7}{6}$. *Shift*, *SD* and the answer should be $1\frac{1}{6}$

Example 04:

$$\frac{3}{4} - \frac{5}{8} =$$

Calculator: FB 3↓4→-FB 5↓8= $\frac{1}{8}$

Example 05:

$$\frac{2}{3} \times \frac{7}{8} =$$

Calculator: FB 2↓3→xFB 7↓8= $\frac{7}{12}$

Example 06:

$$\frac{3}{7} \div \frac{2}{3} =$$

Calculator: FB 3↓7→/FB 2↓3= $\frac{9}{14}$

1.11 *Using a calculator*

(Whole numbers with fractions and Decimals)

Where whole numbers with a fraction are part of the calculation it can be entered using the *shift* then *FB* button.

Example 07:

$$1\frac{3}{4} + 2\frac{2}{3} \times \frac{7}{8} - 2\frac{3}{8} =$$

Calculator: shift FB 1→3↓4→+ shift FB 2→2↓3→x FB 7↓8→- shift FB 2→3↓8= $\frac{41}{24}$

Now press *shift*, *SD* and the answer should be $1\frac{17}{24}$

Checking the calculations by hand: Firstly convert the denominators of each of the fractions. The lowest common denominator (lowest number) will be 24.

3 x 8 = 24.

If $\frac{3}{4}$ is converted to 24ths the denominator has to be multiplied by 6. Therefore the numerator must also be multiplied by 6. (6 x 4 = 24 and 6 x 3 = 18). $\frac{3}{4} = \frac{18}{24}$

The next fraction $\frac{2}{3}$ must be converted to 24ths and would become $\frac{16}{24}$. (8 x 3 = 24 and 8 x 2 = 16).

$\frac{7}{8}$ converted to 24ths becomes $\frac{21}{24}$ (3 x 8 = 24 and 3 x 7 = 21).

$\frac{3}{8}$ converted to 24ths becomes $\frac{9}{24}$ (3 x 8 = 24 and 3 x 3 = 9).

The calculation becomes $1\frac{18}{24} + 2\frac{16}{24}$ x $\frac{21}{24} - 2\frac{9}{24}$ =

VERY IMPORTANT: The order that a calculation is carried out is dictated by the acronym: BIDMAS. Before carrying out the calculation see the section on BIDMAS.

1.12 How to convert decimals to fractions using a calculator

The denominator of a fraction is the number of parts of 1 (whole number). For example $\frac{1}{2}$ states there are 2 equal parts (denominator) and there is 1 of them (numerator).

$\frac{3}{4}$ states that there are 4 equal parts of 1 and that there are 3 of them.

$\frac{7}{8}$ shows there arc 8 cqual parts and there are 7 of them, and so on.

To convert the fraction into a decimal divide the denominator into 1 and multiply by the numerator.

$\frac{1}{2}$ = The calculation will be 1 ÷ 2 = 0.5

0.5 x 1= 0.5.

Example 08:

$\frac{3}{4}$ The calculation will be 1 ÷ 4 = 0.25

0.25 x 3 = 0.75

Example 09:

$\frac{7}{8}$ The calculation will be 1 ÷ 8 = 0.125

0.125 x 7 = 0.875

Another way of writing it as one calculation is: (1 ÷ 8) x 7 = 0.875

1.13 *How to convert fractions to decimals using a calculator*

Use the SD button (Standard to Decimal). Standard will be fractions. Convert the previous examples into decimals.

Example 08:

$\frac{3}{4}$

Calculator: FB 3↓4=$\frac{3}{4}$, SD and the result will show 0.75

Example 09:

$\frac{7}{8}$

Calculator: FB 7↓8=$\frac{7}{8}$, SD and the result will show 0.875

1.14 *Solving problems*

Problems are rarely written in mathematical terms other than in text books. Therefore in real life the first stage is to establish what is required and how it can be solved.

Example 10:

An architectural technician has been given a set of original plans for a house that were drawn up in imperial units. (Imperial units mean; feet and inches, pounds and ounces, gallons and pints).

The alterations and extension to the house will be drawn up in metric units; metres and millimetres, kilograms and grams, litres and millilitres.

The imperial units will need to be converted into decimal units.

a) 12' 9"

b) 10' 3"

c) 6' $5\frac{3}{4}$ "

d) 3lb 4oz

e) 13lb 5oz

f) $1\frac{3}{4}$ galls

g) $3\frac{2}{3}$ pts

To be able to convert the units the following can be used:

12' (' denotes feet). 9" (" denotes inches). There are 12 inches in 1 foot (feet is the plural of foot).

1" = 25.4mm, therefore 1' must equal 304.5mm. As half a millimetre is so small and in practical terms a bricklayer cannot work to such a small measurement 305mm is generally used.

3lb (lb denotes pounds in weight). 4oz (oz denotes ounces). There are 16oz in 1lb. There are 28.35 grams in 1 ounce.

1gall (gall denotes gallon). 3pts (pts denotes pints). There are 8 pints in 1 gallon or 4.546lts. 1pint equals 568.261ml (millilitres).

Text books and Websites don't normally explain the units. It is assumed that the reader will know what they are. However if you were born in or after the 1970s, or in the mainland Europe, you will probably have never used imperial units.

The USA currently uses feet and inches, and pounds and ounces however American gallons are not the same volume as British gallons. Likewise the other liquid volumes such as the pint are also different.

On inspection;

a) 12' 9" could be considered as $12\frac{3}{4}$ feet.

9" is $\frac{3}{4}$ of a foot.

Using a calculator enter: shift FB 12→3↓4→x305=SD and the result will show 3888.75. The answer will be in millimetres therefore convert to metres by moving the decimal point to the left 3 places. The result will be 3.889m.

That was using the rounded up figure of millimetres per foot (305). Now compare if millimetre per inch had been used.

The calculation will be 12'x12" (the result will be the number of inches), add 9" and multiply by the number of millimetres per inch.

VERY IMPORTANT; Note the data has to be entered into the calculator in a specific way otherwise the result will be incorrect.

Entering the data as follows: 12x12+9x25.4=SD will produce a result of 372.6 which obviously is not correct. However if the same data is entered as: (12x12+9)x25.4=SD the result will be correct 3886.2. Why the difference?

The objective is to convert the feet and inches to inches, then multiply the result by the conversion factor of 25.4. In maths terms brackets would be put around the figures that have to be calculated first. (12x12+9) and then x25.4. For more detail see BIDMAS.

There are two methods of entering the data into the calculator: 12x12+9=x25.4=SD and the result will be 3886.2. (3.886m). Or by using the bracket buttons: (12x12+9)x25.4=SD. Both methods provide the correct answer.

The second method is better though as later when more complex calculations are carried out using brackets will be much easier.

There is also another consideration. By using the rounded up conversion figure for feet to millimetres an error of 3mm has occurred over the length.

b) Now compare the length 10' 3".

Using a calculator enter: 10x12+3x25.4=SD and the result will be 196.2 which is obviously incorrect.

Now enter the same data but using brackets.

(10x12+3)x25.4=SD and the answer should be 3124.2. (3.124m).

c) 6' 5$\frac{3}{4}$ "

Convert the fraction to a decimal and then complete the calculation.

4 into 1 = 0.25 and 3 of them; 0.25 x 3 = 0.75"

6x12+5.75x25.4=SD and the answer will be 218.05 which is incorrect.

Now using the brackets; (6x12+5.75)x25.4=SD and the correct answer is 1974.85mm. Or 1.975m.

d) Requires converting 3lb 4oz to metric weights. The first stage is to convert everything into the same units. At the moment there are pounds and ounces.

Convert the pounds to ounces and then into grams.

3x16+4x28.35=SD and the answer 161.4 which is incorrect.

(3x16+4)x28.35=SD and the correct answer is 1474.2g, or 1.474kg.

Following the rules of BIDMAS the calculation within the brackets will be completed before anything else.

Without the brackets the numbers will be multiplied together first and the results added together: 3x16=48 and 4x28.35=113.4. Then the results would be added together: 48+113.4 = 161.4 (Incorrect).

Using BIDMAS brackets first: (3x16+4)=52. Then multiply the result by 28.35.   52x28.35=1474.2 (Correct result)

It is essential to include the brackets around the numbers that need to be calculated before being influenced by another number.

e) 13lb 5oz. (13x16+5)x28.35=SD and the correct answer is 6038.55g.

As there are 1000 grams in a kilogram move the decimal point 3 places to the left. It is normal to round up or down the very small measurements therefore the answer will be 6.038kg.

Always check what units the answer is to be shown in especially in exam questions.

f) $1\frac{3}{4}$ galls. Convert the gallons into pints and then into millilitres.

Using the calculator enter the number and fraction as follows:

(shift FB 1→3↓4→x8)x568.261= and the correct answer is: 7955.654ml. That can also be shown as 7.956ltrs.

g) $3\frac{2}{3}$ pts. Using the calculator enter;

(shift FB 3→2↓3)x568.261= and the correct answer is: 2083.624ml. That can also be shown as 2.084ltrs.

2.00 Brackets and BIDMAS

What happens when there are several calculations that need to be carried out?
For example 2 x 3 + 7 − 6 ÷ 2 =

Do you work from left to right? From right to left, or the easiest bit first? There is a simple rule that states the order of all calculations; BIDMAS or you may have been told BODMAS if it was years ago.

Brackets

Indices

Division

Multiplication

Addition

Subtraction

We will stick to using 'BIDMAS'.

As there are no brackets or indices (we will come to those shortly) the next one is **divide** (÷). It can also be shown as a forward slash (/).

$2 \times 3 + 7 - 6 \div 2 =$

Therefore $6 \div 2 = 3$, or $6 / 2 = 3$

$2 \times 3 + 7 - 3 =$

Multiplication is next; $2 \times 3 = 6$

$6 + 7 - 3 =$

Addition is next; $6 + 7 = 13$

$13 - 3 = 10$

Check it with your calculator. Enter $2 \times 3 + 7 - 6 \div 2 =$

What do these symbols mean? ( ) , [ ], { }

They are 3 different types of bracket. The bandy legs ( ) brackets are technically parenthesis brackets. In math's they can be used to include a particular calculation that should be carried out before any other calculations hence it is the first letter of BIDMAS.

If there are several blocks of calculations each with their own parentheses (pair of brackets) they can be enclosed in a second set of parenthesise or a set of square brackets.

$(2 \times 5) + (6 \times 2) - (3 + 1) \times 6$

If we then want it divided by 2 it would become;

$((2 \times 5) + (6 \times 2) - (3 + 1) \times 6)/2$

Or

$$\frac{((2 \times 5) + (6 \times 2) - (3 + 1) \times 6)}{2}$$

If we wanted to then multiply the answer by 4 putting another set of parenthesise brackets would look messy and confusing therefore use a square bracket to enclose the calculation.

$$\left[\frac{((2 \times 5) + (6 \times 2) - (3 \div 1) \times 6)}{2}\right] \times 4$$

2.01 *Scientific notation*

Where very large or very small numbers need to be written in a formula, or just written down, scientific notation can be used. The idea is to show only one unit multiplied by 10 to the power of. (termed the **exponent**).

The power (exponent) indicates how many times 10 should be multiplied by. Using the base of 10 enables easy manipulation as to the number of decimal places either as a positive or negative.

If there are 6 zeros before the decimal point then the exponent will be 6.

4,000,000 becomes 4×10^6.

Written out fully the calculation would be 4x10x10x10x10x10x10.

If the number is 6 places behind the decimal point then the exponent will be 10^{-6}. Note it is to position of the number compared with the decimal point and not how many zeros.

0.000004 becomes $4x10^{-6}$

Examples include very large numbers used in global energy calculations, and very small numbers as used in thermal expansion of materials.

Using scientific notation could reduce the accuracy however it depends upon the requirements of the calculation.

4,010,050J could appear as $4.01x10^6$J. That means 50Joules have been left out of the calculation. It may be insignificant when considering such a large number.

Scientific notation would be used in preference to using larger units in specific types of calculations. In the above example 4,010,050J could also be written as 4.01MJ (Mega Joules).

It is easy to complete the calculations using the calculator. The exponent button $\boxed{x10^x}$

Example 01

Enter 80x4200 and the correct answer will be 336000. Now enter 3.36EB5= and the result should also be 336000.

(To save space EB means exponent button).

Entering very large numbers, or the result becomes a very large number the calculator will automatically present the result in scientific notation.

Enter 15000000000 and the screen should display it as 1.5×10^{10}. In contrast entering very small numbers the calculator will convert the result into negative scientific notation. Try entering; 0.0000000125=SD and the result should be 1.25×10^{-8}.

Example 02

How much wood is needed to bring 1 ltr of water to the boil?

Given information:

1. Specific heat capacity of water is: 4200J/kg/K.

[Joules per kilogram of water per Kelvin. Where Joules are measurement of energy, kilogram is the mass of 1 litre of water, and Kelvin is the unit of temperature measurement].

2. Calorific value of dry wood: 15GJ/t = 1.5×10^{10}J/t

[GJ is Giga Joule where Giga represents 1,000,000,000, and t represents tonnes, metric unit of weight]

3. Density of the wood: 600kg/m^3

[Where density = mass/volume].

In calculation water has a presumed temperature of 20°C unless otherwise stated. The mass of water; 1kg/ltr or 1g/ml is true at 20°C at sea level.

So far specific data has been provided that includes very large and very small units of measurement.

Stage 01 [How much heat energy is required]

Calculate how much heat energy is required to raise the temperature of the water from 20°C to 100°C. Kelvins are the same sized increments as used in the Celsius scale therefore the temperature difference will be 80K. (note it isn't degrees Kelvin, just Kelvin)

The amount of heat energy required to raise the temperature of water is the SHC (Specific Heat Capacity). For an increase of 1Kelvin 4200J will be required.

$80 \times 4200 = 336000J = 336kJ = 3.36 \times 10^5 J$

Stage 02 [Calculate the heat energy produced by burning wood]

It is very unlikely that a cubic metre of wood will be required therefore convert the unit size to something smaller; cubic centimetres (cm^3).

There are 100cm in 1m therefore 1,000,000 cm^3 in $1m^3$. 1×10^6 cm^3 in $1m^3$ ($100 \times 100 \times 100$).

To reduce the calorific value unit size from tonnes to kilograms divide by 1000. 15GJ/t divided by 1000. (1 tonne =1000kgs).

15,000,000,000J per tonne/1000=15,000,000J/kg or 15MJ/kg or $1.5 \times 10^{10} J/kg$

[The type of units such as Joules, kg or Kelvins etc. would not be shown in the calculation, only in the result. However to enable easier following they have been shown for reference only]

Now the main parts of the calculation are in comparative units of kilograms:

- Heat value of the wood is in MJ/kg
- Density of the wood is in kg/m^3
- Specific Heat Capacity of water is in MJ/kg/K

The calculation of the heat energy released from burning wood is next.

Calorific value of wood x the density of 1m^3 of wood then converted to cm^3.

Based on 1cm^3 of wood = 15MJx600kg/m^{3}x10^{-6}MJ=9kJ.

It can also be written as; $1.5 \times 10^7 \times 600 \times 10^{-6} = 9 \times 10^3$J

The heat energy required has been calculated as 336kJ. If 1cm^3 of wood produces 9kJ then divide 336/9=37cm^3.

This could also be written as; $(3.36 \times 10^5)/(9 \times 10^3) = 37$cm^3.

2.02 Powers (Index numbers)

Smaller numbers written to the right of a number indicate 'power' or index to the number. Basically how many times the number is to be <u>multiplied by itself</u>. Where more than one index number is involved the plural is indices.

To confuse issues it is also known as an 'exponent' – see Exponents. From that word is derived the word exponential

The second command of BIDMAS is to use any indices – see BIDMAS for further detail.

Indices can also be used with letters used in formulae such as a^2, ab^{12}, y^{-3}. The use of letters is known as algebra.

Some examples involving numbers:

5^2 means 5x5=25. This is referred to as 'squaring a number'.

5^3 mean 5x5x5=125. This is termed 'cubed'. Five cubed.

4^6 means 4x4x4x4x4x4x4=4096.

There are several different buttons on the calculator that will complete the calculation without having to put all the information in long hand.

The squaring button

Type in 5 and then press the x^2 button and equals. On the screen 5^2 will appear top left and the result 25 will be bottom right.

The button is used to power the number by 3

(cubed: $x \times x \times x = x^3$)

Type in 5 and then press the x^3 button and equals. On the screen 5^3 will appear top left and the result 125 will be bottom right.

A third button enables any power (indices) to be entered ($x^\square$).

Type in 4 and press the $x^\square$ indices button then 6 and =. Top left on the screen will appear 4^6 and bottom right the result 4096.

Example

(4x4)+(7x7x7)+(8x8x8x8x8x8x8)= the result is 262503

Stage 01 is to complete the contents of each set of brackets.

16+343+262144=262503

The information can also be entered into the calculator using the square, cubed and indices buttons. $4x^2+7x^3+8x^\square6=262503$.

Where the power is negative then use the  after entering the $x^\square$.

Note the negative button has brackets around it and not to be confused with the minus button that has no brackets

To enter 4^{-10} enter $4x^\square(-)10$=SD and the result will be $9.536743164 \times 10^{-7}$. Top left of the screen will display 4^{-10} and bottom right will be the result.

There isn't a negative square or cube and the calculator will display syntax error. There is a way around it but it is easier to enter the data via the $x^{\square}$ button as previously mentioned.

Indices can be multiplied or divided. For example 3^5 means 3 multiplied by itself 5 times: 3x3x3x3x3. If 3^5 is then multiplied by 3^2 the result would be 2187. (3x3x3x3x3)x(3x3) which is the same as 3^7. To multiply the rule is add the indices only if the number or letter is the same.

Division would entail subtraction of the indices. $3^5 \div 3^2$ means (3x3x3x3x3) $\div$ (3x3) = the result is 27. Completing the data in the brackets first: 243 $\div$ 9 = 27.

If you enter the data in the calculator then the direction button must be used otherwise the wrong power will be entered therefore the incorrect result.

Example $3^5 \times 3^2$

Enter $3x^{\square}5 \rightarrow x3x^2$ = the result is 2187.

If you get $2.954312707 \times 10^{21}$ you didn't press the direction key on the large round button.

Example $3^5 \div 3^2$

Enter $3x^{\square}5 \rightarrow \div 3x^2$ = the result is 27.

Number to a power can also be 'raised to a power'.

Example

If 5^5 is to be raised to the power of 3 it would written $(5^5)^3$. This would be the same as (5x5x5x5x5)x(5x5x5x5x5)x(5x5x5x5x5). As can be seen it is much easier and neater to use indices than long hand. The result is 3.052×10^{10}. Calculate the data within the brackets first: $3125 \times 3125 \times 3125 = 3.052 \times 10^{10}$

Check it using the calculator. Enter $(5x^\square5\rightarrow)$ $x^\square3 = 3.052 \times 10^{10}$. It is essential to enter the direction button before closing the bracket otherwise the result will be incorrect. Look at the screen. If the data top left reads $(5^{5)}$ then the direction button has not been used.

2.03 *Roots (Index numbers)*

Where power or index numbers (indices) indicate how many times a number is multiplied by itself, roots are the opposite. It indicates what multiplied by itself becomes the number.

7^2 is 49. Therefore the root of 49 is 7. That particular root is the square root. To find the square root of a number using a calculator press the square root button enter the number, then =.

To find the cube root of a number use the shift button then the square root button(SR), enter the number and then =.

Example

To find the cube root of 64.

Shift SR 64 = and the result is 4. On the screen; top left display a small 3 then the root symbol of a tick with a bar, and 64 below. The result will be bottom right of the screen.

The calculator can be used to find the square root or cubed root of any number including decimal places.

Example

To find the cube root of 614.35.

Shift SR 614.35 = the result is 8.501. To check your answer press AnsxAnsxAns=SD

Where another root (termed 'nth root) is to be found press the shift button followed by $x^{\square}$. Top left of the screen a small empty square to the left of the root sign. Enter the required root, then the direction button and the base number.

Example

To find the fifth root of 64.35.

Shift$x^{\square}$5→64.35= the result 2.9999. To check the result press AnsxAnsxAnsxAnsxAns=SD.

2.04 Exponents

Exponent is another term used instead of power or indices. It is actually different though. Where a number has 'to the power of' it indicates how many times that number should be multiplied by itself. An exponent enables an unknown figure to be shown.

For example 5^3 means 5x5x5 however there may be occasion where it is not known how many times the base number (in this example 5) has to be multiplied. The exponent would be 5^x where the x indicates a variable figure.

From that word is derived the words exponential and exponentially.

2.05 Logarithms

Long before calculators every student had a log book full of codes. It enabled very large or small numbers to be multiplied or divided more easily by using logarithms and anti-logarithms. But what is a logarithm?

It is basically a coded way of writing a short calculation. For example; 5x5x5x5=625. It could be shown as $\log_5(625)=4$.

The code shows that the base number (the number that is being multiplied by itself. 5 in this example) as $\log_5$. Then in brackets the result of the multiplication which is 625. The result equals the number of times the number has been multiplied.

Example

What does the code $\log_3(729)=?$

It means how many times does the number 3 need to be multiplied by itself to equal 729. The result is 6.

3x3x3x3x3x3=729= 6 times

But how can it be calculated? Use the button

I have used LSO (Log Square Oblong) to mean that particular button.

Press LSO button, then 3 followed by direction button to the right, then 729 = and the result is 6.

LSO3→729= The screen will show $\log_3(729)$ top left and the result 6 bottom right.

2.06 Common logarithms

In construction and Civil Engineering a common base of 10 is used. This is known as the Common Logarithm. There is a specific button on the calculator that automatically uses base 10 **Log**

Press log then enter 1000, close bracket and press =. The result is shown bottom right on the screen as 3.

Log1000)= and the result is 3

2.07 *Natural logarithms*

Another useful logarithm is the Natural Logarithm where the base number is unknown. The lower case letter 'e' is used as the base standing for Euler's Number. (Euler was an 18th century Swiss physicist who developed many of the modern day math's).

Euler's number for base 'e' is about 2.71828. On the calculator the button marked [In]

Example

2.71828x2.71828=7.389 approximately. The symbol used to show that the result is an approximation is a wavy equals sign ≈.

Using the calculator enter In. The first open bracket also appears on the screen. Then enter 7.389 and close the bracket and press = and the result is almost 2.

Enter the data using the LSO button. LSO2.71828→7.389= and the result will be almost 2.

Chapter 3 – Profit margin

3.00 *Percentages*

Per cent literally means per hundred. The term is also used as part of a whole. For example 'I am one hundred percent sure'. In mathematical terms one hundred percent is often written as 100%. It can also be shown as 1/100 indicating one hundredth of the number.

The numerator (the number above the line) can range from 1 to 100 showing the percentage. For example; $\frac{25}{100}$ the fraction can be reduced by dividing the numerator into the denominator. 25% will become $\frac{1}{4}$

(Divide the numerator by 25 = 1 and the denominator = 4).

So if there is 25% of the pizza left, 'of' means multiply as in '3 boxes of screws'. Therefore 25% of the pizza means 1x $\frac{25}{100}$

As previously mentioned the fraction can be reduced and becomes $\frac{1}{4}$ of the pizza left.

Decimals can also be used to describe a percentage as in 33.33%. Using the pizza analogy again if there is 33.33% of the pizza then the calculation could be written as 1x $\frac{33.33}{100}$ and the result would be 0.3333.

Using the calculator enter 1x33.33 shift (=SD. The top left of the screen will display 1x33.33% with 0.3333 bottom right as the result. Pressing the shift button and then the '(' button calculates percentages. It operates the yellow symbol % which is above the '(' button.

Whole numbers, fractional numbers or decimal numbers can all be used to indicate a percentage.

Example

a) 59% of £2500
b) 49.5% of £100,000
c) 3000W with a 95.5% efficiency
d) £385 with a 17½% discount

Using a calculator enter

a) 59 ÷ 100 x 2500 = and the result is 1475.
Alternatively enter the same data as:

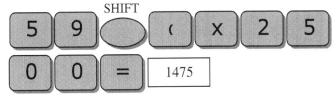

b) 49.5x100000 shift (= and the result is 49500.
Alternatively it can be written as 49.5x100000÷100=

c) 3000x95.5 shift (= and the result is 2865. Alternatively it can be written as 3000x95.5÷100=

d) 385xshift FB 17→1↓2→ shift (= SD and the result is 67.375. (FB means the fraction button). Alternatively it can be re-written as 385x17.5÷100=

3.01 *Problems*

Why have different ways of writing the same thing? Problems are rarely shown as mathematical calculations other than in a text book. The biggest issue is to extract what is required and then re-write it in a math's format.

Calculations need to be clearly shown. Apart from it makes it easier to complete it also helps anyone else who needs to check them. In an exam there are marks awarded for showing the various stages of a calculation.

Calculations and formulae used in math's and science often contain a horizontal line with numbers and or letters both above and below the line as in fig 3.01.

$$\frac{49.5 \times 2500}{100} = 1237.5$$

Fig 3.01

The line indicates that anything above the line will be divided by anything below the line.

Cancelling

Cancelling can help reduce the size of the calculation making it clearer. The 2500 above the line could be reduced to 25. The math's calc would be:

$2500 \div 100 = 25$

Cancelling can be used when the same letters appear above and below the line – fig 3.02.

$$\frac{49.5 \times 25\cancel{00}}{1\cancel{00}} = 1237.5$$

$$\frac{2a + \cancel{b}}{\cancel{a} + \cancel{b}} = a$$

Fig 3.02

3.02 *One number as a percentage of another.*

It's quite a common problem. To find out what a profit margin was, or what discount percentage has been offered.

Example 01

A discount of £51.38 off the total of £685 has been given. What is the percentage discount?

The information shows £685 as the total and £51.38 discount. So the question is; what is £51.38 as a percentage of £685?

Divide the amount of discount by the total sum and then multiply by 100.

51.38 ÷ 685 x 100 = SD and the result is 7.5%. To check the answer is correct enter 685 x 7.5 shift (= SD.

Example 02

The permissible wastage for face brickwork is 8%. How many extra bricks should be ordered if 45,300 bricks are required?

45300x8 shift (= and the result is 3624 extra bricks.

Example 03

The surveyor has measured the number of bricks laid part way through the job (3600 bricks) and compared it with the amount delivered so far (4320 bricks). What is the percentage wastage?

Stage 01.

Subtract the number of bricks laid from the total delivered. That is the amount of wastage. 4320-3600= the result is 720.

Stage 02.

Divide the wastage by the total delivered and multiply by 100.

$720 \div 4320 \times 100 = SD$ and the result is 16.67%

3.03 *How to calculate fractional percentages*

Example 04

Percentages are easier when working with decimals therefore convert fractions to decimals. To convert any fraction to a decimal divide the denominator (number below the line) into the numerator (number above the line).

There are two alternative ways to use the calculator; Using the fractions button or calculate by dividing the bottom number into the top number (changing the fraction into a decimal)

Example 05

Using the calculator calculate 10% of $\frac{5}{8}$

Enter $\frac{5}{8}$ x10 shift (=SD and the result will be 0.0625.

If you require the result to be shown as a fraction then there is no need to press SD. The result will be $\frac{1}{16}$

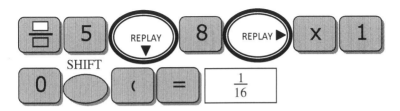

3.04 *How to convert a fraction into a percentage*

Example 06

Using the calculator convert $\frac{43}{50}$ into a percentage.

The question is actually; What is 43 as a percentage of 50?

$43 \div 50$ x100 = and the result is 86%

43 is 86% of 50. To check it is correct type: 50x86shift(=

3.05 *How to convert a decimal percentage into a fraction*

A decimal percentage 0.75% has been shown on an estimate. To compare it with another it must be converted to a fraction.

Method

Using a calculator enter 0.75shift(= and the result will be $\frac{3}{400}$ which is $\frac{3}{4}$ of 1%.

3.06 *Ratios and percentages*

Example 07

A freshly plastered wall requires sealing before decoration. The label on the PVA suggests two dilution rates; 1 part PVA to 10parts water, or 10%. How much PVA needs to be measured out?

The coverage rate is 15m² per litre of diluted PVA. Two coats are required. The wall area is 32m²

Dilution in this case is based on parts by volume. The volume mentioned is litres.

Method 1 – Using the ratio

Calculate the volume of diluted PVA required.

Wall area 32m² x 2 coats = 64m².

Coverage rate is 15m² per litre therefore divide the area by the coverage rate: 64 ÷ 15 = the result is 4.26ltrs

Using the dilution ratio of 1 in 10

- divide the total by 10 and multiply by 9. The result is the volume of water required.

- Subtract the volume of water from the total volume will show the volume of PVA required: 4.26 – 3.84 = 0.42ltrs

Method 2 – Using the percentage

Calculate the volume of diluted PVA required.

Wall area 32m² x 2 coats = 64m².

Coverage rate is 15m² per litre therefore divide the area by the coverage rate: 64 ÷ 15 = the result is 4.26ltrs

Using the dilution ratio of 10%

- Multiply the total by 10%. The result is the volume of PVA required. The result is 0.43ltrs

[This figure is slightly less accurate as it is based on both PVA and water combined]

3.07 Can a percentage be greater than 100%?

In short; kind of. 100% is a comparative figure meaning in comparison to the whole. It is also a figure of speech though as technically a percentage cannot exceed 100.

An example of a figure of speech; The sales figures were 110% up on last month. Meaning they are actually 10% more than last month's figure.

It cannot be 110% of the tree is used with no waste though. The most that can be used is 100%.

Figures of speech are often misnomers. Another common one in the industry is 'heigth'. As in 'length, width and heigth'.

"When a tree is felled it can contain 130% water". It cannot be true. If the tree was 100% water it would mean there was no wood, just water.

So what is the comparison is indicating? It is the weight of the water in comparison to the weight of the timber if it was dry.

Example 08

A tree as felled contains natural moisture and weighs 1.3tonne and when dry weighs 1 tonne.

By calculation a percentage moisture content can be found:

$$\frac{\text{Wet weight} - \text{dry weight}}{\text{Dry weight}} \quad \text{x} \quad \frac{100}{1} \quad = \quad \% \text{ mc}$$

Using the calculator enter:

((1.3-1) ÷1)x100=   The correct result is 30%mc.

If the brackets had not been used and the data entered as follows the result would be incorrect:

1.3-1÷1x100 = An obvious incorrect result of -98.7

It is essential to enter the data correctly.

There are other calculations where a percentage greater than 100% can be correctly used. To add VAT at 20% to a total is easier multiplying by 120% than multiplying by 20% and then adding it to the total.

Example 09

The goods and services total: £46.25 + VAT (20% at time of writing)

46.25x20Shift(+46.25=SD and the result is £55.50

Or 46.25x120shift(=SD and the result is £55.50

Example 10

Copper prices fluctuate. The supplier has advised there is a 5% increase next month. The quoted price for pipes and fittings is £439 plus VAT at 20%. How much should the invoice be and compare it to ordering next month.

The increase in cost is purely on the material and not the tax, therefore calculate the increase and then add the VAT.

Stage 01

Calculate the total cost including VAT for the original invoice.

439x120shift(=SD and the result is 526.80. That is the total including VAT for the original invoice.

Stage 02

Calculate the price increase for the material.

439x105shift (=SD and the result is 460.95.

By entering the whole cost and percentage increase (100+5=105) the calculator has added the increase.

Stage 03

Ansx120shift (=SD and the result is 553.14.

Stage 04

Subtract the smaller number from the greater number to find the actual increased cost. 553.14 – 526.80 = 26.34.

The invoice will be £26.34 more next month.

All of the above stages could be completed in one calculation though:

((439x105shift()x120shift()-(439x120shift()=SD

However there is more chance of error. By calculating in stages it should be possible to inspect the result after each calculation. Any significant error should stand out.

For example at stage 03, if the result was 55314 it would be obvious the percentage button had not been pressed.

Chapter 4 – Algebra "It's all a lot of mumbo jumbo to me"

4.00 *Algebra*

'What use is it? I don't understand what all those letters meant so I switched off'. Sounds familiar?

Algebra is a tool that enables the unknown to be written down. Instead of writing 'unknown' a letter or symbol can be used. For example 3x?=27. But what happens if there are two unknowns?

Using a question mark is not that useful therefore it is usual to use lower case letters or ancient Greek letters to represent unknowns.

The reason for using lower case letters is to prevent confusion with upper case letters used as constants. The same applies to ancient Greek letters. Generally it is the lower case versions that are used.

A constant is something that is given a letter to save writing down a lot of information. π for example is the ratio between the radius and circumference of a circle.

'Hang on I thought you said constants were upper case?'
There are exceptions

The following lists show both the lower and the upper case ancient Greek letters:

	Upper case	Lower case
Alpha	A	α
Beta	B	β
Gamma	Γ	γ
Delta	Δ	δ
Epsilon	E	ε
Zeta	Z	ζ
Eta	H	η
Theta	Θ	θ
Iota	I	ι
Kappa	K	κ
Lambda	Λ	λ
Mu	M	μ
Nu	N	ν
Xi	Ξ	ξ
Omicron	O	o
Pi	Π	π
Rho	P	ρ
Sigma	Σ	σ
Tau	T	τ
Upsilon	Y	υ
Phi	Φ	φ
Chi	X	χ
Psi	Ψ	ψ
Omega	Ω	ω

Using letters in a formula enables unknowns to be written in a statement or calculation.

Example 01

a – 2 = 4. 'a' is the unknown quantity. It is clearly indicated as opposed to using the letter 'x' that could be mistaken to indicate multiplication. Therefore certain lower and upper case letters are not used (i, l, o and x).

4.01 *Simple algebra*

There are rules that must be obeyed and an order for the procedure, but the actual calculations are the same as for numbers; addition, subtraction, multiplication, and division.

Symbols are given to save writing the words: + - x ÷ .

Letters can be shown to the power of in the same way as numbers. For example with numbers; 2^2 means 2 x 2, or with letters; a^2 means a x a.

Brackets enclose data that should be resolved before it is used elsewhere. For example (9-5) ÷ 3 = . If the data in the brackets has not been calculated first then the result will be 7.333 which is incorrect. Completing the data in the brackets and then dividing by 3 will provide the correct result of 1.33.

The same applies if the data is letters or other symbols and that is where algebra becomes useful.

There are general rules that apply to any math's calculations;.

- Keep all the calculations neat.

- Where possible the equals sign should be kept in line vertically throughout – see fig 4.01.

- any mistakes should either be erased or strike a diagonal line through with the correction next to it (don't try over-writing as the end result may not be obvious later and becomes a problem)

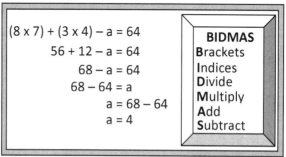

$(8 \times 7) + (3 \times 4) - a = 64$
$56 + 12 - a = 64$
$68 - a = 64$
$68 - 64 = a$
$a = 68 - 64$
$a = 4$

BIDMAS
Brackets
Indices
Divide
Multiply
Add
Subtract

Fig 4.01

4.02 *Functions using algebra*

Adding and subtracting in algebra. Where the same letters appear then any function can apply.

Where a number is shown in front of a letter it indicates multiplication in the same way as saying 'I have 4 apples'.

The following statement can be simplified.

4a + 7b + 2a – 4b

Using the car analogy 'a' (red cars) and 'b' (blue cars).

There are 4 lots of red cars (4a) plus 2 lots of red cars (2a) therefore they can be arranged next to each other. Plus 7 lots of blue cars (7b) and 4 lots of blue cars (-4b) to be taken away.

4a + 2a + 7b – 4b

Simplified it becomes 6a + 3b (6 red cars and 3 blue cars). Letters can be used instead of writing the whole statement.

Where possible simplify more complex statements – see section 4.06.

4.03 *Transposition*

A word used by mathematicians to mean 're-arrange'. Transpose the formula means re-arrange it. However it also could mean to complete parts of the formula during the process.

Generally writing a statement of what is known and what is unknown involves an equals sign (=).

Where possible the unknown should be one side (ideally to the left) of the equals sign and the known to the right side. The unknown would be termed the 'subject'.

The equals sign is like a wall. If a number or symbol (+) is on the left side of the wall and it is moved across it becomes (-). Vice versa applies.

What is actually happening is to move a number or letter from the left side the opposite has to be done to cancel it.

Example 01

12 + 4 = 16

To move the 4 from the left side a minus 4 must be added to both sides. On the left side it cancels it out. + 4 -4 = 0   see fig 4.02.

On the right side of the equals sign it becomes just minus 4.

12 + 4 -4 = 16 -4

12 + 0 =12

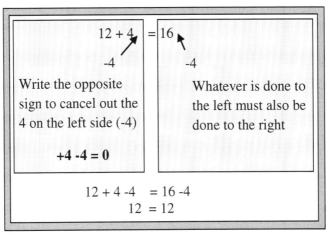

Fig 4.02

Exactly the same principle works for using letters – fig 4.03

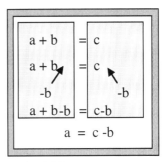

Fig 4.03

Example 02

$$a + e = f + d$$

Becomes

$$a + e - d = f$$

To prove the transposition is correct substitute numbers for the letters – fig 4.04.

Comparing using
numbers:
$$12 + 6 = 9 + 9$$
$$12 + 6 - 9 = 9$$

Fig 4.04

It is good practice to show the subject on the left side therefore the whole equation is shown as a mirror image. The signs do not change. (An equation is a statement that has an (=) sign where something is equal to something else).

$$f = a + e - d$$

4.04 *Formulae and set relationships*

Formulae are statements with set relationships. They can be written out in full or letters used to represent the words. For example the area of a circle can be calculated using a set formula.

The statement is a constant of 3.142 times the radius squared will equal the area. Yes long winded. So set letters are used instead;

3.142 becomes 'π', the radius becomes 'r' with a superscript [2] indicating the radius is to be multiplied by itself.

$$\pi \, r^2 = A$$

Relationships involving sides of a triangle are shown against a circle. Commonly known as trigonometry the long winded versions for right angled triangles have been reduced to three set formulae – see Trigonometry.

One of the formulae states; the tangent angle is equal to the length of the side opposite that angle divided by the length of the line adjacent to that angle. Shorter versions are more commonly used.

$$\text{Tan } \alpha = \frac{\text{Opposite}}{\text{Adjacent}} \qquad\qquad \text{Tan } \alpha = \frac{O}{A}$$

Using the formula on the right enables easy and clear transposition. It is very important to use the recognised letters for the specific words though.

The following formula is used to calculate density.

$$D = \frac{m}{v}$$

This formula enables the density of gas to be calculated;

$$\frac{p_2 V_2}{T_2} = \frac{p_1 V_1}{T_1}$$

However unless the reader is very familiar with the short codes just writing a formula can present problems. Writing a key on the sheet is the solution:

V = volume, p = pressure and T = temperature.

The density formula; 'D' is density, 'm' is mass and 'v' is volume.

Note the use of upper and lower case letters as they have different meanings. Upper case K represents Kelvin when calculating temperatures whereas lower case k represents thermal conductivity, or 1000 as in kilogram, kilometre etc.

[λ lambda is the current symbol for thermal conductivity]

4.05 *Transposition above and below the line*

There are rules to transposition. Consider the '=' sign as a scaffold board on a wall. If a number or letter on one side is to be transferred to the other side the board must remain balanced.

Diagram 'A' in fig 4.05 shows the density formula. To illustrate the process numbers have been substituted for the letters – dia. 'B'. $3 = 12 \div 4$.

If the number above the line doesn't have anything below the line it is a whole number or unit. It is the same as writing a 1 under the line.

$3 \div 1$ is still 3. Likewise the rule applies to any other whole number or unit.

Dia. 'C' shows $4 \div 1$.

If the 12 and the 1 change places the equation is still balanced – dia. 'D' and 'E'.

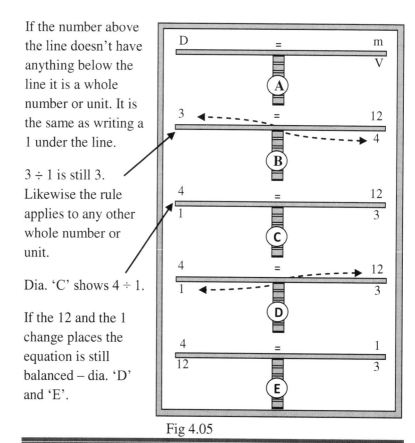

Fig 4.05

The rule is diagonals can be swapped without changing signage.

4.06 *Equations involving numbers and letters*

Equations involving number and letters can be complex involving calculations on one or both sides of the equals sign.

Example 03

Transpose and simplify $\dfrac{36x^2y^3z^5}{4x^2yz^3}$ =

Stage 01

Isolate the numbers and complete the calculation.

Stage 02

Isolate the letters one at a time that appear above and below the line and complete the calculations.

Stage 03

Gather the results – see fig 4.06

$$\frac{36x^2y^3z^5}{4x^2yz^3} = 9 \times 1y^3z^2$$

$$\frac{36}{4} = 9$$

$$\frac{x^2}{x^2} = 1 \qquad \frac{x \times x}{x \times x} = 1$$

$$\frac{y^3}{y} = y^3 \qquad \frac{y \times y \times y}{y} = y^3$$

$$\frac{z^5}{z^3} = z^2 \qquad \frac{z \times z \times z \times z \times z}{z \times z \times z} = z^2$$

Fig 4.06

Note: dividing x^2 by x^2 is not x. It is 1. Substituting numbers for the letters is a good way of checking the transposition is correct.

$x = 3$, $y = 4$ and $z = 5$

Stage 01

Isolate the numbers and complete the calculation.

Stage 02

Isolate each substituted letter one at a time that appear above and below the line and complete the calculations.

Stage 03

$$\frac{36x3^2 x4^3 x5^5}{4x3^2 x4x5^3} = 9 \times 1 \times 16 \times 25$$

$$3600 = 3600$$

$$\frac{36}{4} = 9$$

$$\frac{3^2}{3^2} = 1 \qquad \frac{3 \times 3}{3 \times 3} = 1$$

$$\frac{4^3}{4} = 4^2 \qquad \frac{4 \times 4 \times 4}{4} = 4^2$$

$$\frac{5^5}{5^3} = 5^2 \qquad \frac{5 \times 5 \times 5 \times 5 \times 5}{5 \times 5 \times 5} = 5^2$$

Fig 4.07

Gather the results – see fig 4.07

Yes the two results agree so must be correct. It takes more time but worth checking as the result may not always be correct.

5.00 *Regular areas*

Calculating areas

Area of a:

Square: Area = l x b

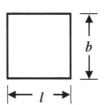

Rectangle: Area = l x b

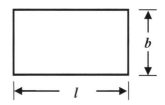

Parallelogram; Area = l x h

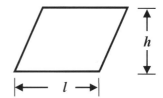

Trapezium: Area = (a + b) / 2 x h

$$\text{Area} = \frac{(a + b)}{2} \times h$$

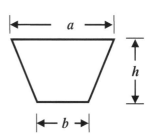

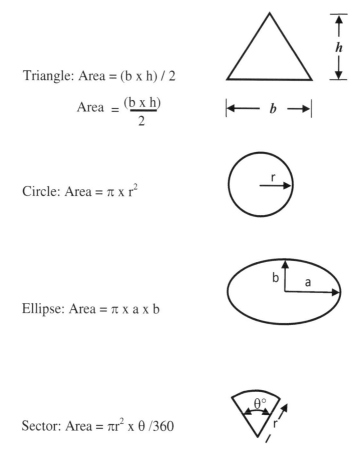

Triangle: Area = (b x h) / 2

Area $= \dfrac{(b \, x \, h)}{2}$

Circle: Area = π x r^2

Ellipse: Area = π x a x b

Sector: Area = πr^2 x θ /360

Using the above formulae should enable most flat regular areas to be calculated. Irregular areas require more involved processes many of which were developed by Thomas Simpson in the 18th century.

5.01 *Irregular areas [Simpson's Rule]*

Simpson's rule and the mid-ordinate rule are based on a similar idea. Both use the area of a trapezium.

Example 5.01

The shape in fig 5.01 has one irregular side and three straight sides. Sides 'A' and 'B' can be measured easily and an average calculated to give line 'C'.

In this example C = 4.5m – fig 5.02.

If the width of the rectangle is 2m and it is multiplied by the length of line 'C' then an approximate area will be calculated.

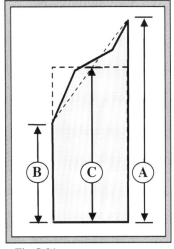

Fig 5.01

The maths calculations are:

$$\text{Area} \approx \frac{(A + B)}{2} \text{ x w}$$

Where 'w' is the width of the strip

$$\text{Area} \approx \frac{(6 + 3)}{2} \text{ x } 2$$

Note the area is only approximate therefore the equals sign changes to 'approximately equal to' $\approx$

$$\text{Area} \approx 4.5 \text{ x } 2$$

$$\text{Area} \approx 9\text{m}^2$$

Depending upon how accurate the area needs to be will dictate how wide the strip width is.

The lines 'A' and 'B' are termed 'ordinates'. Basically it means a straight line at right angles to another straight line.

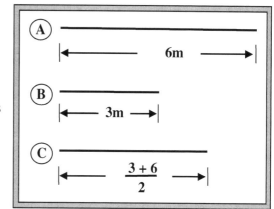

Maps are often divided into squares and the measurements from those lines are referred to as co-ordinates.

Back to the plot.

Fig 5.02

If the area of a large irregular shape is to be calculated divide it up into an equal number of equal width strips.

Example 5.02

Calculate the approximate area of the plot of land shown in fig 5.03.

1. Draw a straight line at right angles to the road at the widest part of the shape. Measure its length.

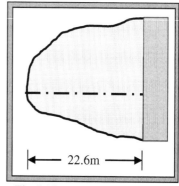

Fig 5.03

2. Divide the shape into an equal number of equal width strips. There are two methods; Use a calculator, or use a scale rule as shown in fig 5.04 and 5.05.

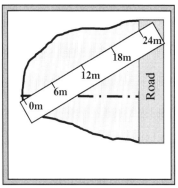

Fig 5.04

3. Using the ruler method: Place the zero at the start and rotate the ruler until an even number is in line with the other side of the shape.

4. Number the end of each ordinate for identification.

5. Measure the length of each ordinate (start from the straight line that goes across the shape). Write the length on each ordinate.

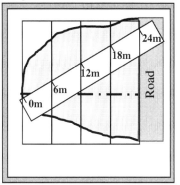

Fig 5.05

Putting a circle / oval round the number will help prevent confusion later – fig 5.06.

6. Note the first ordinate in this example will be zero.

The width of each strip is 5.65m.

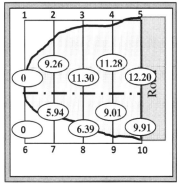

Fig 5.06

5.02 *Using the mid-ordinate rule*

Formula: Area ≈ Width of strip x the sum of the mid-ordinates

It may be abbreviated to: A ≈ w x (Σ mid-ordinates)

'Σ' is Sigma and means 'the sum of'. You will use it when working with spreadsheets.

Example 5.03

Using the data from the example add the length of each pair of ordinates and divide by 2.

Ordinate 1 is zero and ordinate 2 is 9.26.

The calculation will be: (0 + 9.26) / 2 = 4.63(Mid-ordinate 1).

Next pair will be 9.26 and 11.30.

(9.26 + 11.30) / 2 = 10.28 (Mid-ordinate 2).

(11.30 + 11.28) / 2 = 11.29 (Mid-ordinate 3).

(11.28 + 12.20) / 2 = 11.74 (Mid-ordinate 4).

Then do the same for below the line.

(0 + 5.94) / 2 = 2.97 (Mid-ordinate 5).

(5.94 + 6.39) / 2 = 6.165. Round it up to; 6.17 (Mid-ordinate 6).

(6.39 + 9.01) / 2 = 7.70 (Mid-ordinate 7).

(9.01 + 9.91) / 2 = 9.46 (Mid-ordinate 8).

Now each of the mid-ordinates has been calculated (basically an average of the ordinates from each side of the strip) enter the info into the formula.

$A \approx w \times (\Sigma \text{ mid-ordinates})$

> Always write the formula first. It enables focus and allows for easier checking.

$A \approx 5.65 \times (4.63+10.28+11.29+11.74+2.97+6.17+7.70+9.46)$

BIDMAS therefore complete the calc's in the brackets first.

$A \approx 5.65 \times 64.24$

$A \approx 362.956\text{m}^2$. As previously mentioned the result is only approximate therefore approximately equals symbol should indicate that.

In most cases the area would be rounded up to whole units therefore 363m^2.

Checking is very important. A rough 'ball park figure' should provide enough comparison. In this case add the ordinates for 5 and 10; $12.20 + 9.91 = 22.11$ and multiply by the width of the plot:

$22.11 \times 22.60 = 500$. Allow a bit for the irregular shape and 363m^2 appears to be correct.

5.03 *Using the Trapezoidal rule*

Look on the Internet at the various web-pages and the explanations of how to use the trapezoidal rule. Why do they make it so complicated?

This is a simplified version of the Trapezoidal rule.

$$\text{Area} \approx w \left(\left(\frac{\Sigma \ 1^{st} + \text{last}}{2} \right) + \left(\Sigma \text{ of all the other ordinates} \right) \right)$$

Where w is the width of each strip and the 1st, last and other ordinates refers to the ordinates as measured. Not the mid-ordinates.

Example 5.04

Using the same example enter the data and calculate – fig 5.06.

Formula first

$$\text{Area} \approx w \left(\left(\frac{\Sigma \ 1^{st} + \text{last}}{2} \right) + \left(\Sigma \text{ of all the other ordinates} \right) \right)$$

Calculate the top area first

$$\text{Area} \approx w \left(\left(\frac{0 + 12.20}{2} \right) + \left(9.26 + 11.30 + 11.28 \right) \right)$$

Area ≈ 5.65 x (6.10 + 31.84)

Area ≈ 5.65 x 37.94

Area ≈ 214.361m²

> Double brackets in an equation. The rule is to complete the calculation within the inner brackets first. Then the outer brackets.

Calculate the bottom area and add to the top area.

$$\text{Area} \approx w \left(\left(\frac{\Sigma\ 1^{st} + \text{last}}{2} \right) + \left(\Sigma \text{ of all the other ordinates} \right) \right)$$

$$\text{Area} \approx w \left(\left(\frac{0 + 9.91}{2} \right) + \left(5.94 + 6.39 + 9.01 \right) \right)$$

Area ≈ 5.65 x (4.96 + 21.34)

Area ≈ 5.65 x 26.30

Total area ≈ 148.595m^2 + 214.361 = 362.956m^2. Round up to whole units becomes 363m^2.

Thomas Simpson developed a more accurate method in the 18th century. It still is an approximation albeit more accurate.

5.04 *Using Simpson's Rule*

$$\text{Area} \approx \frac{w}{3} \left((1^{st} + \text{last}) + 4(\text{evens}) + 2(\text{odds}) \right)$$

Where

'w'	=	width of strip
'1st'		the first ordinate
'last'		the last ordinate
'evens'		the even numbered ordinates
'odds'		the odd numbered ordinates

Note they are ordinates and not mid-ordinates.

Example 5.05

The process of dividing the irregular area is the same as for the previous methods therefore use the same data from the previous example.

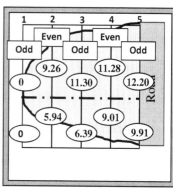

Fig 5.07

Identify each set of ordinates as either odd or even – fig 5.07.

Complete the top area first then complete the lower area.

Formula first

$$\text{Area} \approx \frac{\text{w}}{3}\left((1^{st} + \text{last}) + 4(\text{evens}) + 2(\text{odds})\right)$$

Now enter the data

$$\text{Area} \approx \frac{5.65}{3}\left((0 + 12.20) + 4(9.26+11.28) + 2(11.30)\right)$$

$$\text{Area} \approx \frac{5.65}{3}\left(12.20 + 4(20.54) + 2(11.30)\right)$$

$$\text{Area} \approx \frac{5.65}{3}\left(12.20 + 82.16 + 22.60\right)$$

$$\text{Area} \approx \frac{5.65}{3} \text{ x } 116.96$$

$$\text{Area} \approx 1.883 \text{ x } 116.96$$

$$\text{Area} \approx 220.275$$

BIDMAS
Brackets
Indices
Divide
Multiply
Add
Subtract

Now complete the bottom area.

$$\text{Area} \approx \frac{w}{3}\bigg((1^{st} + \text{last}) + 4(\text{evens}) + 2(\text{odds})\bigg)$$

$$\text{Area} \approx \frac{5.65}{3}\bigg((0 + 9.91) + 4(5.94+9.01) + 2(6.39)\bigg)$$

$$\text{Area} \approx \frac{5.65}{3}\bigg(9.91+ 4(14.95) + 2(6.39)\bigg)$$

$$\text{Area} \approx \frac{5.65}{3}\bigg(9.91 + 59.80 + 12.78\bigg)$$

$$\text{Area} \approx \frac{5.65}{3} \text{ x } 82.49$$

$$\text{Area} \approx 1.883 \text{ x } 82.49$$

$$\text{Area} \approx 155.329$$

Top area 220.275 + bottom area 155.329 $\approx 376\text{m}^2$

Comparing the:

mid-ordinate rule 363m^2

trapezoidal rule 363m^2

Simpson's rule 376m^2

$$\frac{13}{363} \text{ x } 100 = 3.58\%$$

$$\frac{13}{376} \text{ x } 100 = 3.46\%$$

There is a variance of 13m^2. As a percentage the result using Simpson's rule can be expressed as 3.58% more than the mid-ordinate and trapezoidal rule or they are 3.46% less than using Simpson's rule.

5.05 *Areas of parts of a circle*

Areas of parts of a circle such as sectors, segments and quadrants are based on the ratio of the radius against a constant represented by pi (π).

The constant is the number of times the diameter will go into the circumference of the circle – fig 5.08.

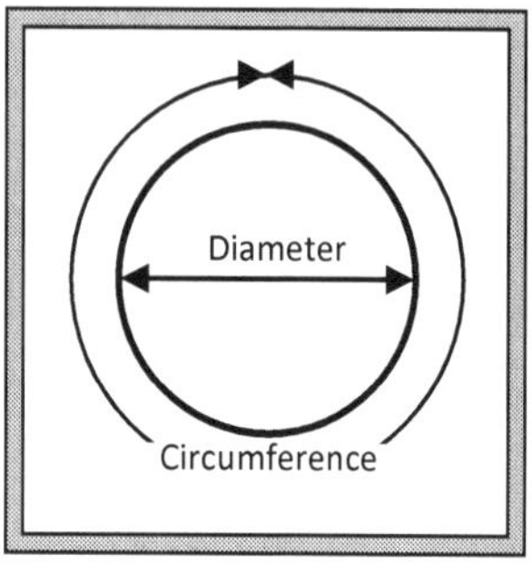

The result is always:

Circumference = πd.

It can also be written as $2\pi r$

Fig 5.08

For general calculations π = 3.142 or 22/7.

Example 5.06

If the diameter of a circular rod is 15mm what will be the length of the circumference?

Formula first: Circumference = πd

Circumference = 3.142 x 15

Circumference = 47.12mm

Using the calculator there is a button specifically for π. It is accessed by pressing shift then the exponent button.

Using a calculator press the buttons in the order shown below:

SHIFT

Area of a circle

To calculate the area of a circle the formula required is πr^2

Area = πr^2 where 'r' indicates the radius.

Example 5.07

What is the cross sectional area of a circular rod 15mm in diameter?

Stage 01 is to convert the data from diameter to a radius.

Formula: d/2 = r

15/2 = 7.5mm

r = 7.5mm

Write down the area formula:

Area = πr^2

A = 3.142 x 7.5^2

$A = 176.72\text{mm}^2$

Using the calculator:

Area of a sector

Formula: Area = πr^2 x θ /360

A slice of pie is a sector. There have to be two known pieces of data;

1.   the radius
2.   the internal angle

In the diagram the internal angle is shown as lower case theta (θ) representing any angle up to 359°.

Example 5.08

What is the area of a sector with a radius of 3m and internal angle of 23°.

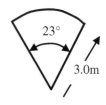

Formula first

Area of sector = πr^2 x θ /360

Enter the data

Area = π x 3^2 x 23 / 360

Area = $1.806m^2$

An alternative method would be to calculate the area of the whole circle, divide by 360 and multiply by 23.

5.06 *Area of an ellipse*

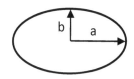

Formula: Area = π x a x b

An ellipse can be considered as either a stretched or squashed circle with an additional radius termed axis. The longer radius is the major axis and the shorter the minor axis.

Example 5.09

To calculate the weight of an elliptical handrail the cross sectional area must be calculated and then multiplied by the length of the rail. Axis: a = 35mm and b = 25mm.

Formula: Area = π x a x b

Enter the known data: A = π x 35 x 25

Using the calculator:

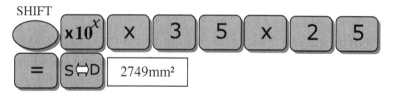

To carry out a quick rough check: A = πr² using the larger axis.

$A = \pi r^2$

$\mathbf{A} = \pi \times 35^2$

A = 3849. Therefore 2749 appears to be correct.

5.07 *Calculating the major and minor axes*

Example 5.10

A rainwater pipe 75mm dia is to go through a pitched soffit board in a roof. The pitch (angle from horizontal) is 40°. Calculate the major and minor axis and the inclined surface area – fig 5.09.

The minor axis will be r = d/2.

$r = 75 / 2$

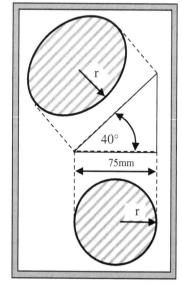

Fig 5.09

r = 37.5mm

The major axis can be calculated using trigonometry – fig 5.10.

(SOH, TOA, CAH) See Trigonometry for more detail.

Select a suitable formula. Only one has hypotenuse and adjacent therefore it must be cosine.

Formula first

$$\text{Cos } \alpha = \frac{\text{Adj}}{\text{Hyp}}$$

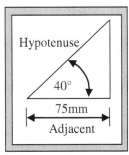

Transpose the formula before entering any data.

$$\text{Hyp} = \frac{\text{Adj}}{\text{Cos } \alpha}$$

Enter the data and calculate

Fig 5.10

$$\text{Hyp} = \frac{75}{\text{Cos } 40°}$$

Using the calculator

Hyp = 75 / Cos 40

Hyp = 97.91

Divide the hypotenuse by 2 and that will provide the length of the major axis.

Major axis = 97.91 / 2

Major axis = 48.95mm (there is a difference of 0.01 in accuracy if using the calculator as it is using 8 decimal places. For the purposes of cutting a whole 49mm would be accurate enough).

To calculate the inclined surface area just means the area of the ellipse.

Formula first

Area = π x a x b

Enter the data

Area = 3.142 x 37.5 x 49.0

Area = 5773mm^2

Using the calculator

Carry out a rough check

A = πr^2 using the major axis as 'r'

A = π x 49 x 49

A = 7543mm^2 therefore the area of the ellipse appears to be correct as it should be slightly less than a circle.

Using the calculator

SHIFT

If the display shows: 2401π it means the SD button wasn't used.

5.08 *Area of a segment*

A segment is an area of a circle between a chord and the circumference. A chord is any straight line between two points on the circumference of a circle - fig 5.11. The diameter is the longest chord possible.

The radius of a circle is any line from the centre to the circumference – fig 5.11.

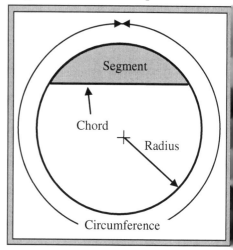

A sector is an area between two radii. Fig 5.12 shows that a sector can be divided into a segment plus a triangle.

The internal angle of the triangle shown as θ can be divided by 2 to form two right angled triangles.

Fig 5.11

The small square top right hand corner indicates it is 90°.

The hypotenuse of the triangle is also the radius of the circle. The area of one triangle can be flipped over to form a rectangle.

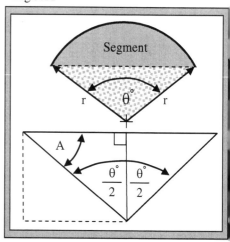

Fig 5.12

5.09 *Alternative formula for the area of a segment*

Area of segment $= \dfrac{1}{2} \; r^2 \left(\dfrac{\pi}{180°} \; \theta - \text{Sine } \theta \right)$

To try out the formula use the data from example 5.10.

The formula can only be used if both the radius and internal angle of the sector are known. The example 5.10 is a method that can be used if the internal angle is unknown.

Known data: radius = 54.450mm. Internal angle 80°

Area of segment $= \dfrac{1}{2} \; r^2 \left(\dfrac{\pi}{180} \; \theta - \text{Sine } \theta \right)$

Area of segment $= \dfrac{1}{2} \; 54.450^2 \left(\dfrac{\pi}{180} \; 80 - \text{Sine } 80 \right)$

Area of segment $= 609.942\text{m}^2$

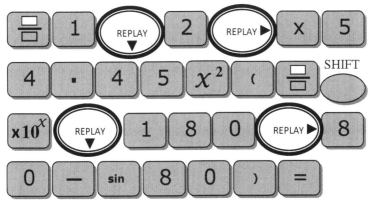

Example 5.10

A paved area to the front of a
building is a segment on plan.
(Looking down from above)

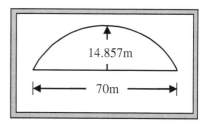

The frontage of the building is
70m and the offset 14.857m.

Fig 5.13

The arc radius is 54.450m – fig 5.13.

(An arc is the part of the circumference that borders the chord –
the curved line).

What is the surface area of the segment?

From fig 5.14 and fig 5.15 two sides
of a triangle are noted. The opposite
is half the length of the chord (70 / 2
= 35), and the hypotenuse is the
radius. Using trigonometry calculate
angle α.

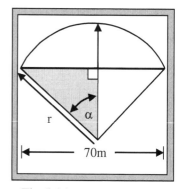

Fig 5.14

Select a suitable formula.

Writing down the pneumonic will
help (SOH, TOA, CAH). There is
only one set with both the opposite
and hypotenuse in it – Sine.

$$\text{Sine } \alpha = \frac{\text{Opp}}{\text{Hyp}}$$

Fig 5.15 shows the right angled triangle that forms half the area of to be calculated. The other half of the triangle is a mirror image of the first – fig 5.16.

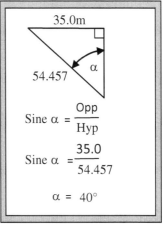

Fig 5.15

Using the 35.0m as the opposite the adjacent can be calculated. It is then a simple rectangle; area = l x b.

Area = length x breadth

A = 35.000 x 41.711

A = 1459.90m²

Next step is to calculate the area of the sector. The internal angle will be 2 x α = 80° (2 x 40°). The given radius is 54.450m.

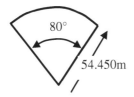

(Note α and θ just mean an unknown angle. θ is used in most text books as part of the formula for the area of a sector).

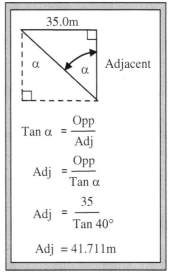

Fig 5.16

Formula:

Area = πr^2 x θ /360

Area = π x 54.450^2 x 80 /360

Area of sector = 2069.823m².

Deduct the area of the triangle previously calculated;

Area of segment = area of sector minus area of triangle

A = 2069.82 – 1459.89

Area of segment = 609.93m²

Using the calculator

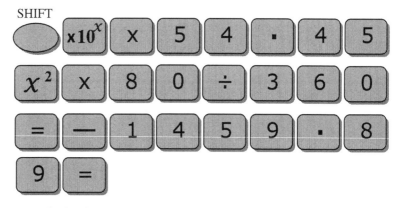

A rough check would be:

35.00 x 14.857 ≈ 520, and 70 x 14.857 ≈1040. Therefore 609.93 appears to be correct.

5.10 *Other area*

Most other areas such as polygons (a shape with many straights sides) can be broken down into rectangles or triangles. As a general rule subtract areas rather than adding them.

Fig 5.17 shows an irregular polygon. If the internal angles and line lengths are provided then it should be easy to calculate a series of right angled triangles and rectangles.

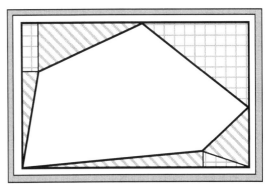

Fig 5.17

Total up their areas and deduct them from the large rectangle.

5.11 *Problems involving areas*

The final section on areas is partly 3 dimensional although still involves an area.

Example 5.11

What is the wetted area per metre of a drainage pipe with an internal diameter of 320mm. The depth of water is 110mm from the invert?

Sketching the problem is useful so that all the information can be seen – fig 5.18.

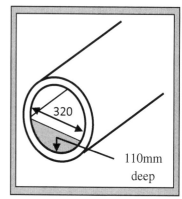

Fig 5.18

Stage 01

Calculate the length of circumference per degree of the circle – fig 5.19.

Formula first

Circumference = πd

C = π x 320

C = 1005

1degree = 1005 / 360

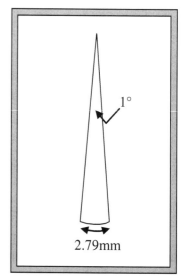

Fig 5.19

1degree = 2.79mm at the circumference.

Stage 02

Calculate the internal angle of the sector.

The water level is 50mm lower than the halfway mark (lower semi-circle) – fig 5.20.

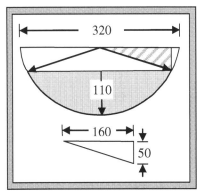

Calculating the angle from horizontal to the top of the liquid can be achieved using trigonometry.

The triangle has an adjacent and opposite therefore use the tangent formula (SOH, TOA, CAH).

Fig 5.20

Formula first

$$\text{Tan } \alpha = \frac{\text{Opp}}{\text{Adj}}$$

$$\text{Tan } \alpha = \frac{50}{160}$$

$$\text{Tan } \alpha = 17.35°$$

To find the internal angle multiply the angle by 2 and deduct from 180°.

Internal angle = 180 - (17.35 x 2)

Internal angle = 145.3°

Stage 03

Final stage is to multiply the number of degrees of the internal angle (145.3°) by the linear on the circumference per degree (2.79mm). That is the length of the arc.

Length of arc = 145.3 x 2.79

Length of arc = 405.387mm

Convert to metres

Length of arc = 0.405387m. That will be the width dimension for the wetted area.

Now multiply by 1m (the length of the wetted pipe area) to convert to an area.

The calculation would be:

Wetted area = (145.3 x 2.79)/1000 x 1

Wetted area = $0.41m^2$

If the units had been left in millimetres the result would be:

Wetted area = 145.3 x 2.79 x1000

Wetted area = $405,387mm^2$

Example 5.12

Using the information from the previous example calculate the number of litres of water that will have flowed in 10minutes based on a flow rate of 1metre per second (1m/sec).

The calculations will be to find the area of the segment (water in the pipe). Then multiply by 1m to produce a volume of water. That will be the amount that passes in 1 second. Then multiply by 10 minutes. That's the plan.

Stage 01

Calculate the sector.

Formula first

Area of sector = πr^2 x θ /360

Enter known data and calculate

Area of sector = π x 160^2 x 145.3 /360

Area of sector = 32,460.33mm^2

Rough check consider the area as a rectangle

Area = Length x breadth

Area = 320 x 160

Area = 51,200mm². Allow a bit for the curves therefore the calculation appears correct.

If the result had been 512,000 then re-check the rough guide before checking the original result.

Stage 02

Calculate the area of the segment.

Method

Sector area less the triangle. See example 5.10.

Area of sector = 32,460.33mm^2 (from stage 01)

The triangle dimensions can be found in fig 5.20.

Area of triangle = (l x h) / 2

Area of triangle = (320 x 50) / 2

Area of triangle = 16,000 / 2

Area of triangle = 8,000mm²

Segment area = Sector area less area of triangle

Segment area = 32,460 - 8000

Segment area = 24,460mm^2

Rough check: Consider the area as a rectangle.

Area = Length x breadth

Area = 320 x 110

Area = 35,200 therefore calculation appears correct. It is only a rough check. Allow say about a third for the curve.

(35,200 x 0.66 = 23,232mm²).

Stage 03

Convert the area into square metres by dividing the area by 1000,000. ($1000,000\text{mm}^2 = 1\text{m}^2$)

A = 24,460 / 1000,000

A = 0.02446m^2

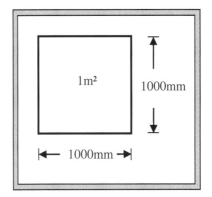

Using the calculator   Fig 5.21

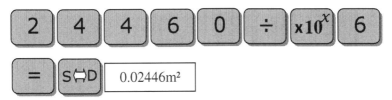

Using the exponent button saves typing in lots of zeros.

Stage 04

Multiply the area by 1m of pipe run will give the volume of water per second in the pipe (volume in m^3)

Volume = 0.02446 x 1

Volume of water per second = 0.02446m^3

Stage 05

Convert the number of minutes into seconds and multiply by the volume of water per second.

Number of seconds = 10 x 60

Number of seconds = 600

Stage 06

Multiply the volume per second by the number of seconds.

Volume of water passing in 10 minutes = 0.02446 x 600

V = 14.676m^3

By converting to m^3 it is an easy conversion to litres. There are 1000ltrs per m^3.

V = 14.676 x 1000

V = 14,676 ltrs

Total volume of water passing through the pipe in 10 minutes is 14,676 ltrs.

6.00 *Calculating volumes (formulae)*

Volume of a:

Cube: Volume = l x b x h

Rectangle: Volume = l x b x h

Cone: Volume $= \pi r^2 \times \dfrac{h}{3}$

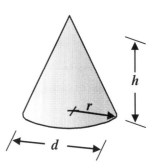

Frustum:

Volume $= \left(\pi r^2 \times \dfrac{(h1 + h2)}{3}\right) - \left(\pi r^2 \times \dfrac{h2}{3}\right)$

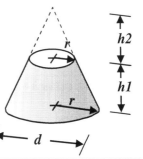

Cylinder: Volume = πr^2 x h

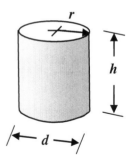

Pyramid: Volume = base x $\dfrac{h}{3}$

The formula works with all shapes of base from triangular through to any polygon.

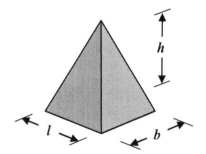

Triangular prism:

Volume = $\dfrac{b \times h}{2}$ x l

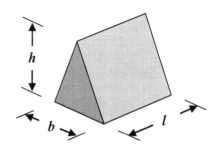

Sphere: Volume $= \dfrac{4}{3}\,\pi r^3$

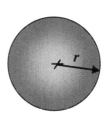

Dome: Volume $= \dfrac{\left(\dfrac{4}{3}\,\pi r^3\right)}{2}$

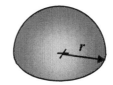

6.01 *Using volumes formulae*

Example 6.01

Calculate the mass of a wall 2.70m long and 2.40m high. The wall is built in concrete blocks (440 x 215 x 100mm) with a density of 2400kg/m^3. The mortar beds are 10mm thick with a mass of 1800kg/m^3. The bond is stretcher bond.

There is a lot of information therefore break it down into stages.

Stage 01

Calculate the mass of 1 block. The density has been given in kg/m^3 therefore convert the dimensions to metre units. (Decimal point 3 places to the left).

Formula:

Volume = l x b x h

V = 440 x 215 x 100

V = 0.440 x 0.215 x 0.100

V = 9.46 x 10^{-3}m^3

Mass = density x volume

$M = 2400 \times 9.46 \times 10^{-3}$

Mass of 1 block = 22.704kg

Stage 02

Calculate the mass of mortar for 1 block.

(bed + a thickness + perp).

To calculate the volume: l x b x h

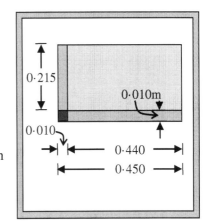

Fig 6.01

Volume = ((bed + mortar thickness) + 1 perp) x 0.1 x 0.01

$V = ((0.440 + 0.010) + 0.215) \times 0.1 \times 0.01$

$V = (0.450 + 0.215) \times 0.01 \times 0.1$

$V = 6.65 \times 10^{-4} \text{m}^3$

Mortar for 1 block = $6.65 \times 10^{-4} \text{m}^3$.

Calculate the mass of the mortar for 1 block

Mass = density x volume (mass of mortar is 1800kg/m^3)

M = 1800 x (6.65 x 10^{-4})

Mass of mortar for 1 block = 1.197kg

Stage 03

Calculate the face area of 1 block including the mortar.

Face area means the surface that would be seen if looking at the wall.

The bed is the layer the block sits on and the perp' (short for perpendicular) is the upright at one end – fig 6.02.

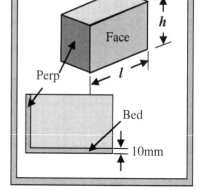

Fig 6.02

Formula:

Area = l x h

Add the thickness of the mortar to the length and then add the height – fig 6.01.

Area = (l + mortar) x h

Area = (0.440 + 0.01) x 0.225

BIDMAS

A = 0.450 x 0.225

Face area of 1 block plus mortar = 0.10125m²

Stage 04

Calculate the area of the wall.

Formula:

Area = l x h

A = 2.70 x 2.40

Area of wall = 6.48m²

Stage 05

It is useful to collect the results from each stage so far:

Stage 01: Mass of 1 block = 22.7kg

Stage 02: Mass of mortar for 1 block = 1.197kg

Stage 03: Face area of 1 block including mortar = 0.10125m²

Stage 04: Area of wall = 6.48m²

The objective is to calculate the mass of the wall therefore calculate how many blocks including mortar would be required for the wall.

Number of block inc. mortar = Wall area / block inc mortar

No. = 6.48 / 0.10125

Number of blocks inc. mortar = 64

Stage 06

The final stage is to multiply the number of blocks by the mass of 1 block and the number of mortars by the mass of 1 mortar.

Mass of blocks = No. of blocks x mass of 1 block

Mass of blocks = 64 x 22.7

Mass of blocks = 1452.8kg

Mass of mortar = No. mortars x mass of 1 mortar

Mass of mortar = 64 x 1.197

Mass of mortar = 76.608kg

Combined mass of wall = mass of blocks + mass of mortar

Combined mass of wall = 1452.8 + 76.608

Total mass of wall = 1529.41kg

To do a very rough check calculate the total volume of the wall and multiply by the mass if it were solid block (A). Then again if it were solid mortar (B). The result should be between the two answers.

'A' = 2.70 x 2.40 x 0.10 x 2400

'A' = 1555.2kg

'B' = 2.70 x 2.40 x 0.10 x 1800

'B' = 1166.4kg

As the result is between answer 'A' and 'B' it suggests it is correct.

6.02 *Accuracy v Precision*

A point of accuracy. Although the calculation should be correct it isn't 100% accurate. There is a difference between accuracy and precision.

Engineers tend to work to precision and builders in most cases to accuracy. So what's the difference?

Throw a dart and it goes in at treble 20. That is accurate (unless you were trying for a treble 5). The dart can be anywhere within the designated area of about 360mm² to be accurate.

In contrast an engineer may work to 0.01mm to enable a tight fit. That is precision.

Building materials vary in both size and mass even for the same material therefore calculating the mass of a wall is not precise. The calculation provides a reasonably accurate guide though.

For the linear measurement of land the nearest 10mm is normally acceptable.

Taking measurements for differences in height 1mm accuracy is normal.

7.00 *Calculating irregular volumes*

Road construction may need cuts and fills to maintain levels.
Cutting through mounds and filling dips. Both require the
calculation of irregular volumes.

Muck clearance of spoil heaps. A guide to the volume of material
is required for pricing, and ordering of plant and labour. Spoil
heaps are an irregular volume.

A few years ago estimators and surveyors would carry out
calculations based on the principles of trapezoidal rule and a series
of plates.

However today surveyors can use GNSS/GIS systems that measure
and calculate volumes or areas using hand held smart antenna and
satellites. The digital data can be plotted and used with graphical
software such as AutoCAD that also can carry out calculations for
areas and volumes.

So why do we need to know how to do it the mathematical way?
Simply there may be occasions where the hi-tech mechanical
methods are not available.

7.01 *Simpson's prismoidal rule*

Method

The principle is based on dividing the irregular volume into an odd number of sections through termed 'plates'. Similar to toast on a rack.

The plates are identified by the letters 'A1' and 'A2' for each end and 'M' for the middle plates.

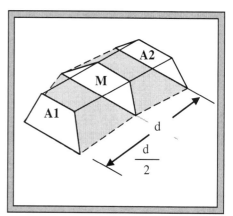

Fig 7.01

For the example only 3 plates have been used however if more plates are required the 'M' plates will also require a number for identification – fig 7.01.

The formula is basically the same as Simpson's rule used for calculating an irregular area. The main difference is the ordinates are now plates.

Formula

$$\text{Area} \approx \frac{w}{3}\Big((1^{st} + \text{last}) + 4(\text{evens}) + 2(\text{odds})\Big)$$

Where

'w'	=	width of strip (d/2)
'1ˢᵗ'		the first plate
'last'		the last plate
'evens'		the even numbered plates
'odds'		the odd numbered plates

Example 01

Three plates have been measured as plate A1, M and A2. Calculate the volume if the overall distance of 40m – fig 7.02.

Stage 01

Calculate area of the plates.

Formula

Area A1 = $\dfrac{(a + b)}{2}$ x h

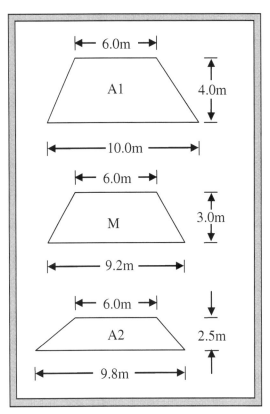

Fig 7.02

Area A1 = $\dfrac{(a + b)}{2}$ x h

Area A1 = $\dfrac{(6 + 10)}{2}$ x 4

Area A1 = 8 x 4

Area A1 = 32m^2

Area M = $\dfrac{(a + b)}{2}$ x h

Area M = $\dfrac{(6 + 9.2)}{2}$ x 3

Area M = 7.6 x 3

Area M = 22.8m^2

Area A2 = $\dfrac{(a + b)}{2}$ x h

Area A2 = $\dfrac{(6 + 9.8)}{2}$ x 2.5

Area A2 = 7.9 x 2.5

Area A2 = 19.75m^2

Stage 02

Calculate w. (w = d/2)

w = 40 / 2

w = 20m

Formula

$$\text{Vol} \approx \frac{w}{3}\Big((1^{st} + \text{last}) + 4(\text{evens}) + 2(\text{odds})\Big)$$

As there is only 1 middle plate it must be an even.

Enter data into formula

$$\text{Vol} \approx \frac{20}{3}\Big((32 + 19.75) + 4(22.8) + 2(\text{zero})\Big)$$

$$\text{Vol} \approx \frac{20}{3}\Big(51.75 + 91.2 + 0\Big)$$

$$\text{Vol} \approx \frac{20}{3} \times 142.95$$

$$\text{Vol} \approx 6.667 \times 142.95$$

$$\text{Vol} \approx 953m^3$$

Volume of irregular shape is about 953m^3. It cannot be very accurate as the volume of the material has been measured at specific points only and it can vary either side of each point.

Carry out a rough check

V = L x b x h

V ≈ 40 x 9 x 3

V ≈ 1080 so the result looks reasonable.

7.02 *Trapezoidal Rule and Mid-ordinate Rule*

In the same way that Simpson's Rule can be applied to volumes so can both the trapezoidal and mid-ordinate rules. It is just the wording of the formulae that changes:

Mid-ordinate rule

Formula: Volume ≈ Width of strip x the sum of the mid-plates

The areas for each plate must be calculated and then given a letter and number for identification. Then apply the formula in the normal way. The mid-ordinate plates are just the average of the plate either side of it.

Example 02

Use the information from example 01 to calculate each mid-plate.

Mid-plate area 1 $= \dfrac{(\text{Plate A1} + \text{Plate M})}{2}$

Mid-plate area 1 $= \dfrac{(32 + 22.8)}{2}$

Mid-plate area 1 $= 27.4\text{m}^2$

Mid-plate area 2 $= \dfrac{(\text{Plate M} + \text{Plate A2})}{2}$

Mid-plate area 2 $= \dfrac{(22.8 + 19.75)}{2}$

Mid-plate area 2 = 21.275m²

Formula: Mid-ordinate rule

Volume ≈ Width of strip x the sum of the mid-plates

V ≈ 20 x (Mid-plate 1 + mid-plate 2)

V ≈ 20 x (27.4 + 21.275)

V ≈ 20 x 48.675

V ≈ 973.5m^3

Trapezoidal rule

$$Volume \approx w \left(\left(\frac{\Sigma \ 1^{st} + last}{2} \right) + \left(\Sigma \text{ of all the other ordinates} \right) \right)$$

$$Volume \approx 20 \left(\left(\frac{32 + 19.75}{2} \right) + 22.8 \right)$$

$$Volume \approx 20 \left(25.875 + 22.8 \right)$$

Volume ≈ 20 x 48.675

Volume ≈ 973.5m^3

The results are less accurate than by using Simpson's Rule. The more cross sections / plates that are used the more accurate the result.

Chapter 8 – Right angled triangles

8.00 *Trigonometry (right angled triangles)*

Basic trigonometry involves the relationships between internal angles and the sides of a right angled triangle.

There are three ratios:

$$\text{Sine } \alpha = \frac{\text{Opposite}}{\text{Hypotenuse}} \qquad \text{Sin } \alpha = \frac{O}{H}$$

$$\text{Cosine } \alpha = \frac{\text{Adjacent}}{\text{Hypotenuse}} \qquad \text{Cos } \alpha = \frac{A}{H}$$

$$\text{Tan } \alpha = \frac{\text{Opposite}}{\text{Adjacent}} \qquad \text{Tan } \alpha = \frac{O}{A}$$

Fig 8.01

Each ratio has been abbreviated into a simple formula. Opposite is short for 'opposite the angle' and adjacent is 'adjacent or next to the angle'.

To make it easier to write, the formulae have been abbreviated down to those shown to the right on fig 8.01.

The ratios are all based on a quadrant of a circle. The hypotenuse must always be the longest side which is the radius of the circle. – fig 8.02 and 8.03.

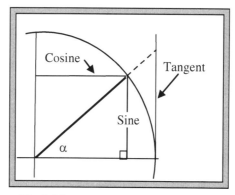

Fig 8.02

The internal angle cannot be greater than 90°.

The adjacent must always be next to the angle – fig 8.03.

The information known and what is required determines which formula is used.

Transposition of the formula (moving it around) should be done whilst it is in the formula format.

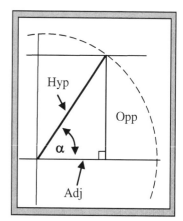

Fig 8.03

In HE and degree level exam's the marking schemes often award marks for writing the correct formula and another mark for correctly transposing before entering the data.

8.01 Which formula should I use?

Always jot down the pneumonic SOH TOA CAH. It will help you select the correct formula. Even in an exam write it down and put a line through it before handing the paper in so it won't be assessed.

SOH, TOA, CAH identifies which sides go with which angle. The following examples show how the formulae can be used.

Example 01

What distance should the foot of the ladder be away from the wall?

Height of the wall is 5.20m to the eaves and the ground is level. Angle is 75° - fig 8.04.

Sketching it out will help.

Objective: to calculate the distance from the wall that the ladder foot should be. (Adjacent)

What is known? The height of the wall and it is opposite the angle.

The angle is 75°.

The two sides are the opposite and adjacent side. Which formula has opposite and adjacent as the named sides? ~~SOH~~, TOA, ~~CAH~~.

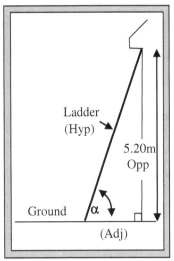

Fig 8.04

$$\text{Tan } \alpha = \frac{O}{A}$$

$$A = \frac{O}{\text{Tan } \alpha}$$

$$A = \frac{5.20}{\text{Tan } 75°}$$

A = Distance from the wall to the foot of the ladder is 1.39m

Compare the ratio of wall to ground (1in 4) to the angle given. The ratio means for every 1m out from the wall the ladder should be 4m up. Multiply the result by 4 and compare with the height.

4 x 1.39 = 5.56m therefore the result is about right.

To check the 1 in 4 ratio the data known is 4 units up (Opposite) divided by 1m out (Adjacent) and inverse Tan of the angle.

It is easier to do the check on the calculator

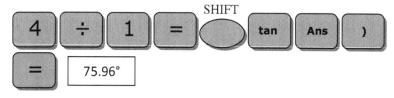

Inverse Tan means the other way round. Instead of using the angle to calculate a side length, inverse function is used to find the angle when the side lengths are known. Accessing the inverse of tan function (tan^{-1}) press the shift button and then tan button.

Example 02

The pitch (angle from horizontal) of a flight of stairs needs to be checked. A simple method is to measure a set distance along the floor and another at right angles vertically – fig 8.04.

Measurements:

Adj = 2.0m

Opp = 1.4m

Formula:

$$\text{Tan } \alpha = \frac{O}{A}$$

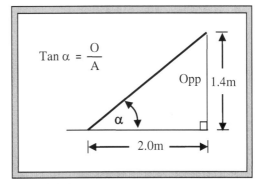

Fig 8.04

Enter the data and calculate.

$$\text{Tan } \alpha = \frac{1.4}{2.0}$$

The pitch of the stairs is 35°.

Using the calculator

8.02 *Angle measurement*

Degrees, minutes and seconds

For many jobs degrees are enough detail in a similar way to hours and time. However smaller increments of a degree are also required.

Why minutes and seconds? Contrary to the many websites and books that state it has nothing to do with time the historical evidence suggests that it does.

The study of astronomy goes back thousands of years. Much of today's geometry was being used in ancient Egypt, Greece and Rome over two thousand years ago.

There are records dating back to the even older civilization of Babylon. In c.1600BC they were using a 360 day year. They based it on the relative position of fixed stars to the Sun being about 1 degree.

The ancient Babylonians had adopted the counting system of the ancient Sumerians who used a base of 60 known as the sexagesimal system c.3000BC.

The navigational instrument the sextant was invented during the 18th century. It is used to sight a known star or the Sun in relation to the horizon at a given time. The measurements are taken in degrees and tenths of a degree using a vernier scale. Not minutes and seconds. This is due to the movement of the vessel preventing exact measurement. It is normal practice to take three readings and average the results.

So back to the plot. Yes a 'degree of arc' as it is known is based on time – being 360 days in the ancient year. Yes the unit of degrees is based on the ancient Sumerian sexagesimal system and would have been divided into 60.

Many centuries later time was divided into groups of 60 (minutes and seconds). Angles and time are related.

There are 360° in a circle. Each degree can be divided into 60 equal parts termed minutes. Each minute can be divided into 60 equal parts termed seconds. That means there are 3600 seconds in a circle (and in an hour).

Decimal degrees

Satellite data that controls the GPS systems uses decimal degrees as opposed to degrees, minutes and seconds. In a similar way to comparing fractions and decimals it is easy to convert degrees, minutes and seconds into decimal degrees using the calculator.

Entering degrees, minutes and seconds on a calculator.

Using the 'tadpole' button

Example 03

To enter the angle 35° 57' 36" (degrees (°), minutes (') and seconds (") press the following buttons:

The screen will display the angle bottom right. To convert the angle to a decimal degree press the tadpole button again.

The same procedure is used whenever angles need to be entered into the calculator. Either enter them as a decimal degree or as degrees, minutes and seconds.

Use the tadpole button to convert between methods.

Example 04

There is a high voltage cable above the ground. The surveyor has gathered data using a total station and pin positioned directly beneath the cable - fig 8.05.

The technique for measuring the height of a building often shown in books will only work if the ground is perfectly level.

[For more information on surveying techniques see Construction Surveying Explained]

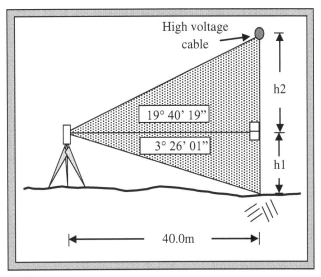

High voltage
cable

h2

19° 40' 19"

3° 26' 01"

h1

40.0m

Fig 8.05

The objective is to calculate the sides 'h1' and 'h2' that are
opposite the angles. The distance is shown as adjacent to the angle.

Select a suitable formula:

~~SOH~~ TOA ~~CAH~~.

Formula

$\text{Tan } \alpha = \dfrac{O}{A}$

$O = \text{Tan } \alpha \times A$

$h1 = \text{Tan } 3° 26' 01'' \times 40$

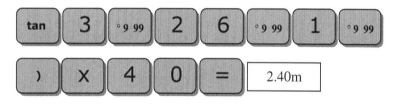

$$\text{Tan } \alpha = \frac{O}{A}$$

$$O = \text{Tan } \alpha \times A$$

h2 = Tan 19° 40′ 19″ x 40

h2 = 14.30

The overall height of the high voltage cable above the ground will be h1 and h2 added together.

Height above the ground: 2.40m + 14.30m = 16.70m

8.03 *Problem solving using trigonometry*

Example 05

Calculating the number of roof tiles for a hipped roof requires the use of trigonometry. Drawings normally show plan view, elevation and section through of the proposed structure.

'Taking off' is the process the estimator will do to gather the information from the drawings. Depending upon the designer the drawings may show the external dimensions of the wall (dashed lines – fig 8.06), or plan dimensions of the roof.

The pitch of the roof (angle from horizontal) is shown in degrees normally on the section through. The other important dimension is the soffit at the eaves of the roof. That is how much the roof overhangs the face of the wall.

Calculating the area for roof tiles is not an exact figure. The roofer will adjust the top courses of the roof tiles to make them fit. However slightly too many is far better than not enough.

Stage 01: Calculate the length of the common rafter.

The common rafter as the name suggests is common throughout the run that meets the ridge board.

Fig 8.06 shows the rafter will be cut at an angle at the top so that it fits flat against the ridge board (Plumb cut). At the bottom of the rafter (the foot) the fascia board will be fixed to the ends of the rafters therefore another plumb cut.

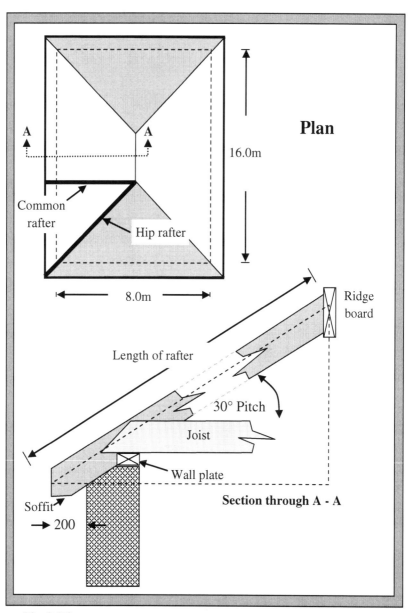

Plan

16.0m

A   A

Common
rafter

Hip rafter

8.0m

Ridge
board

Length of rafter

30° Pitch

Joist

Wall plate

Soffit

Section through A - A

→ 200 ←

Fig 8.06

Stage 01 continued

Calculate the length of the common rafter

Formula (~~SOH~~, ~~TOA~~, CAH)

The angle (pitch) is known; 30°.

The adjacent can be calculated: overall external wall dimension / 2 add the soffit. (8/2) + 0.2 = 4.2m. The rafter is the hypotenuse of the triangle therefore the only formula that includes adj and hyp will be Cosine.

$$\text{Cos } \alpha = \frac{A}{H}$$

Transpose the formula

$$H = \frac{A}{\text{Cos } \alpha}$$

Enter data

$$H = \frac{4.2}{\text{Cos } 30°}$$

Common rafter length is 4.85m

Stage 02: Calculate the length of the hip rafter.

The pitch of the roof is the same for both hipped ends. However the pitch of the hip rafter will be less as it's longer. To calculate how much longer; the common rafter length becomes the opposite and the adjacent will be 4.2m.

Fig 8.07 shows the hip rafter and common rafter on plan view. The distance from the end of the roof to the first common rafter must be exactly the same as half the hipped end plus the soffit overhang.

(8 / 2) + 0.2 = 4.2 Fig 8.06 and 8.07

The hip rafter now becomes the hypotenuse of the new calculation.

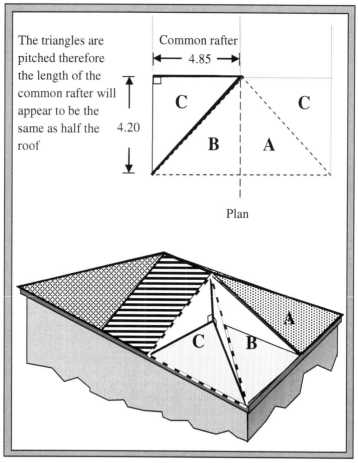

Fig 8.07

The dashed lines show the right angled triangle 'C' – fig 8.07.

Use Pythagoras' theorem; $a^2 + b^2 = c^2$.

'C' can be the hip rafter 'a' the common rafter and 'b' the length of the base line.

Transpose the formula so that the unknown is on the left of the equals sign.

$c^2 = a^2 + b^2$

Enter data

$c^2 = 4.85^2 + 4.20^2$

$c^2 = 23.5225 + 17.64$

$c^2 = 41.1625$

$c = \sqrt{41.1625}$

BIDMAS
Brackets
Indices
Divide
Multiply
Add
Subtract

Hip rafter length is 6.416m

Using the calculator:

Stage 03

The surface area of each hipped end will be triangles 'A' plus 'B'.

There are two simple calc's; either base x common rafter length (that is the easiest); 4.20 x 4.85 = 20.37m²

or area of a triangle:

$$\text{Area of triangle} = \frac{\text{Base x Height}}{2}$$

$$\text{Area of triangle} = \frac{8.40 \times 4.85}{2}$$

Triangles 'C' in fig 8.07 are exactly the same surface area as surfaces 'A' or 'B'. Therefore 2 areas 'C' must equal the hipped end (20.37m²).

Now that all the triangular areas are noted it leaves 2 rectangular shapes as shown in fig 8.07.

The lengths of the sides are the common rafters (4.85m).

The width of the rectangle = the overall length of the roof (16.40m) less the overall width (8.40m).

Sounds complicated but simply put:

Width of rectangle = (16.0 + (2 x 0.20)) − (8.0 + (2 x 0.20)

W = (16.0 + 0.40) − (8.0 + 0.40)

W = 16.40 – 8.40

W = 8.0m

Area of rectangle = w x l

A = 8.0 x 4.85

A = 38.8m²

There are 2 rectangles, one each side therefore:

Total area of rectangles = 38.8 x 2

Total area of rectangles = 77.60m²

Roof tiles

Total up the roof surfaces:

Total roof tile surface = (2 x hipped ends) + (areas ΣC) + (Σ rectangles).   ['Σ' means 'sum of', or added together]

Total roof tile surface = (2 x 20.37) + (2 x 20.37) + (2 x 33.8)

Total roof tile surface = 40.74 + 40.74 + 77.60

Total roof tile surface = 159.08m²

Tile manufacturers normally state how many tiles are required per square metre therefore calculation per tile is not required.

Different types of tile require specials. For example 'plain' tiles require verge tiles and 'a tile and a half' tiles to produce the required lap cover. For more detail look at the tile manufacturer's technical notes commonly found on-line.

Ridge tiles

Ridge tiles will be the 'w' of the rectangles

Ridges tiles = 8.0m

Hip tiles

Hip tiles will be the hip rafter length x 4.

Hip tiles = 6.41 x 4

Hip tiles = 25.64m

Note the ridge tiles and hip tiles are a linear figure and the main surface of the roof tiles in m².

To carry out a quick check the roof surface can be re-arranged so it becomes 1 long rectangle

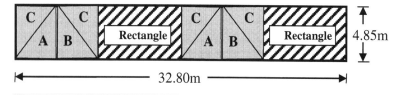

L = (8.0 + (2 x 0.20) + 8.0) x 2

L = (8.0 + 0.40 + 8.0) x 2

L = 16.40 x 2

L = 32.80

Area = l x b

A = 32.80 x 4.85

A = 159.08m²

If plain tiles are being used they require verge tiles and 1½ tiles to enable half lap.

9.00 *Trigonometry (non-right angled)*

There are occasions where right angled triangles cannot be formed so two additional trigonometric formulae can be used; Sine Rule and Cosine Rule.

9.01 *Sine Rule*

Sine Rule: $\dfrac{a}{\text{Sine A}} = \dfrac{b}{\text{Sine B}} = \dfrac{c}{\text{Sine C}}$

The principle behind the formula is the comparison between the internal angle and the side opposite with either of the other two angles and sides.

Example 01

A total station has been set up and readings at stations 'A' and 'B' have been booked.

Linear between stations 'A' and 'B' = 85.5m. Fig 9.01.

'∠' indicates the letter is an angle. ∠A means it is angle A.

∠A = 62.575°

∠B = 56.775°

Two of the three angles are known. Calculate the third

$\angle C = 180° - (62.575° + 56.775°)$

$\angle C = 180° - 119.35°$

$\angle C = 60.65°$

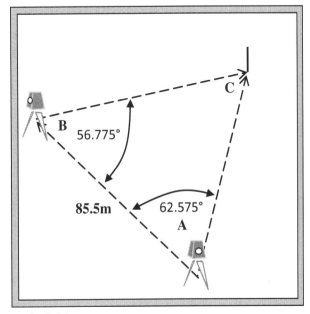

Fig 9.01

Formula

$$\frac{a}{\text{Sine A}} = \frac{b}{\text{Sine B}} = \frac{c}{\text{Sine C}}$$

$$\frac{b}{\text{Sine B}} = \frac{c}{\text{Sine C}} \qquad \textbf{Use part of the formula}$$

Transpose

$$b = \frac{c}{Sine\ C} \times Sine\ B$$

Enter known data

$$b = \frac{85.5}{Sine\ 60.65°} \times Sine\ 56.775°$$

$$b = 98.091 \times Sine\ 56.775°$$

$$b = 82.055m$$

Use the other part of the formula to calculate length of line 'a'.

$$a = \frac{c}{Sine\ C} \times Sine\ A$$

$$a = \frac{85.5}{Sine\ 60.65°} \times Sine\ 62.575°$$

$$a = 98.091 \times Sine\ 62.575°$$

$$a = 87.067m$$

9.02 *Erring on an error*

Carry out a check

Divide the triangle into 2 right angled triangles – fig 9.02.

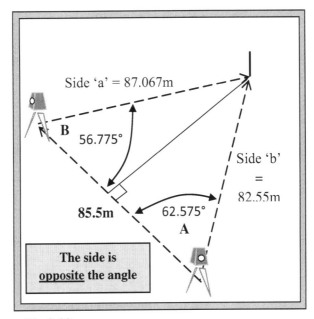

Side 'a' = 87.067m

B 56.775°

Side 'b' = 82.55m

85.5m 62.575° **A**

The side is __opposite__ the angle

Fig 9.02

The hypotenuse and angle are known, the adjacent unknown: (SOH, TOA, CAH).

$$\text{Cosine } \alpha = \frac{\text{Adjacent}}{\text{Hypotenuse}}$$

Transpose

A = Cos α x H

A = Cos A x 82.55

A = Cos 62.575 x 82.55

A = 38.022m

The adjacent for triangle 'A' is 38.022m. *(37.79365631m)**

Now the other triangle

A = Cos α x H

A = Cos B x 87.067

A = Cos 56.775 x 87.067

A = 47.707m

The adjacent for triangle 'B' is 47.707m. *(47.70634369m)**

Add both adjacent lines

38.022 + 47.707 = 85.729m. There is an error of 0.229m.

* denotes all of the decimal places on the calculator had been used, the result is spot on: *(37.79365631 + 47.70634369 = 85.50).*

If the data was being used for setting out say a road or building the inaccuracy would not be acceptable.

Where is the error though? Mainly from rounding up side 'b' from 82.05539056m to 82.055m. The difference rounding up has created is 0.2283436947m, or 228mm.

Rounding up can result in substantial errors therefore using a calculator memory can help if a result is to be used as part of another calculation.

9.03 *Cosine Rule*

The cosine rule is actually three formulae to suit different situations:

$a^2 = b^2 + c^2 - 2bc$ Cosine A

$b^2 = c^2 + a^2 - 2ac$ Cosine B

$c^2 = a^2 + b^2 - 2ab$ Cosine C

A very useful formula when only 1 angle and 2 sides are known.

Example 02

The distance between two posts is to be measured using a total station. A large structure prevents vision between the posts therefore the surveyor has set up the instrument so that both post are in sight – fig 9.03.

The data gathered: angle at station 'A' (58.375°) plus the lineal from station 'A' to each post.

The left post can be named post 'B' and the right post 'C'. The line opposite the angle is given the lower case letter of the angle. The line between the posts becomes line 'a'.

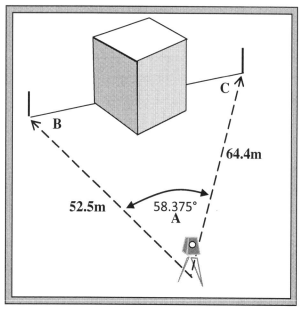

Fig 9.03

The line opposite post 'B' measures 64.4m (b). The line opposite post 'C' measures 52.5m (c).

There is only one Cosine formula that has ∠ A and the unknown side 'a':

$a^2 = b^2 + c^2 - 2bc$ Cosine A

Enter the known data

(Calculate in the order of BIDMAS. Indices first)

$a^2 = 64.4^2 + 52.5^2 - 2$ x 64.4 x 52.5 x Cosine $58.375°$

$a^2 = 4147.36 + 2756.25 - 2$ x 64.4 x 52.5 x Cosine $58.375°$

$a^2 = 4147.36 + 2756.25 - 3545.705361$

$a^2 = 6903.61 - 3545.705361$

$a^2 = 3357.904639$

$a = \sqrt{3357.904639}$

$a = 57.947$m

BIDMAS
Brackets
Indices
Divide
Multiply
Add
Subtract

The distance between the two posts is 57.947m.

9.04 *How to use the memory functions on a calculator*

The calculator has several memory functions. Before using them clear them all by pressing the following buttons:

SHIFT

9.05 *Storing numbers using a memory function*

The Casio fx-85GT Plus has memory stores 'A' through to 'F' plus M+. The older model fx-85ES has 'A' through to 'D' plus M+. Whichever instrument you are using the process is the same.

Calculate 4 x 5 =

To store the result in memory 'A'

To check it is in memory 'A' press

20 should appear bottom right of the screen and 'A' top left.

By pressing 'shift' first then 'RCL' the calculator will use the function 'STO' which means store. (small yellow letters above the button).

Pressing 'RCL' and then the identification button it will display whatever has been stored – fig 9.04. The buttons after 'RCL' are the identification and direct to each memory function.

Entering a second result in the same memory function will delete any existing stored data.

Clear all memories as shown previously.

Check each memory by pressing 'RCL' (recall) and each button as shown – fig 9.04. The memory letter will be top left and '0' bottom right of the screen.

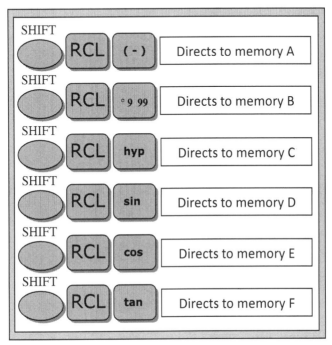

Fig 9.04

9.06 *Using data stored in the memory*

Example 01

There are 2 boxes 4" long. Will they fit into a container 225mm long?

Stage 01

Calculate the overall length of the 2 boxes and store the result in memory 'A'. Then enter 25.4 into memory 'B' (the conversion figure from inches to millimetres).

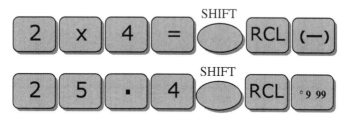

Stage 02

Using the two stored figures multiply them together.

The result is yes the boxes will fit into the container.

There is another memory function that can be used for running calculations. Meaning it is easy to input additional or subtraction to the numbers stored.

9.07 *Using the M+ memory*

Example 02

Clear all the memory functions: shift, 9, 2, = AC

Enter 2 x 5 = . Then press M+ and the screen will display AnsM+ top left and 10 bottom right.

Now enter 2.5 + 7.5 = and press M+ again.

Enter 20 ÷ 2 = and press M+.

The result should be 30 stored in the memory. To check the result press RCL then M+.

Subtracting from the memory can be by adding a negative number and then pressing M+.

Enter 3 x -5 = M+. Check the memory content is 15.

Enter 15M+ to return the memory total to 30.

Now enter 3 x 5 = and press shift M+. By pressing shift the function becomes subtraction from the memory. The shift button is yellow and the M- is also yellow.

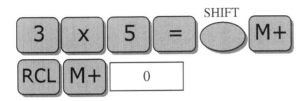

9.08 *Only clearing M+*

To only clear the M+ memory press RCL then M+ to see what is in there and then shift M+ AC. By pressing shift M+ it is effectively adding the negative number of whatever is stored in the memory.

If the button is pressed twice the memory will be the negative of what was stored in there.

Example 03

4 x 5 = then press M+.

Then 20 + 5 = M+.

RCL M+ should display 45.

Now press shift M+ and the AC to clear the screen. The memory will be 0.

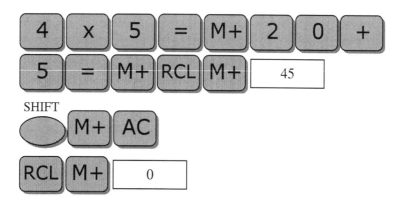

10.00 *Graphs and charts*

These two topics have been put together and include:

- Line graphs
- Straight-line graphs
- Line of best fit graphs
- Bar charts
- Histograms
- Pie charts

The word graph means drawing. It originates from the ancient Greek word 'Graphos'.

'If a picture paints a thousand words ......' is also true of graphs and charts.

10.01 *Line graphs*

Line graphs are most commonly used against two straight lines crossed at right angles. They are the axes (plural of axis). The horizontal axis is identified as the 'x' and the vertical the 'y' axis – fig 10.01.

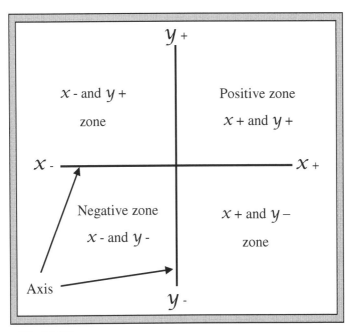

Fig 10.01

Where the two axes cross is normally point zero (although hasn't got to be). To the right will be positive and to the left negative. Vertically up from zero is positive and down is negative – fig 10.01.

That means there are four zones;

$x +$ and $y +$

$x -$ and $y -$

$x +$ and $y -$

$x -$ and $y +$

The x axis is the changeable line (remember nappies – the bottom needs changing).

Example 01

The temperature of a room changes throughout a 24 hour period. Plot a line graph showing the following data.

00.00hrs 15°C, 07:00hrs 18°C, 07:30hrs 22°, 11:00hrs 23°C, 14:00hrs 24°C, 16:00hrs 20°C, 18:00hrs 20°C, 21:00hrs 17°C

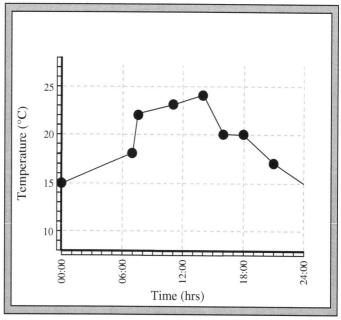

Fig 10.02

As the minimum temperature recorded was 15°C it is useful to show a few degrees less on the graph. It presents the data more clearly – fig 10.02.

Straight line graphs

They are very useful when predicting data from known data.

Water at 20°C has a mass of 1tonne per m³. Using a straight line graph it is easy to plot 1m³ and 1000kg (1tonne). To find the volume of 500kg follow the line horizontally to show 0.50m³.

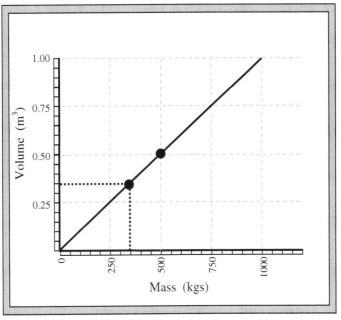

Fig 10.03

To find the mass of 0.27m³ follow the line down to the mass axis and it shows 270kg – fig 10.03.

10.02 *Line of 'best fit' graphs*

Data doesn't always come in neat formats for a straight line graph.
Using a line graph probably won't help either so a 'best fit'
method can be used.

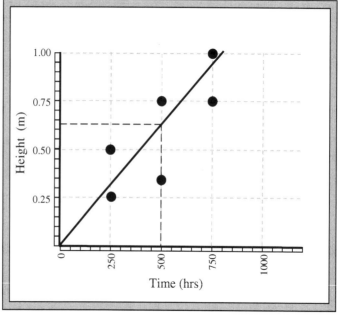

Fig 10.04

The data is plotted on the graph and averaged out so that the
straight line can be drawn in.

Looking at the time axis 500hrs is the average (250 + 500 + 750) /
3 = 500hrs. The average of the heights: (0.25 + 0.50 + 0.75 + 1.00)
/ 4 = 0.625m.

Plot the point where 500hrs and 0.625m meet and draw the straight line through from the zero point. That is the line of best fit. Basically it is an average of the data – fig 10.04.

10.03 *Bar charts*

As the name suggests the data is presented in a series of bars originating from one axis. The width of the bars is aesthetic and has no definition – fig 10.05.

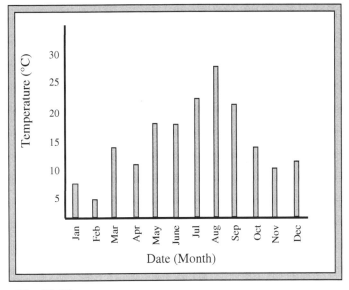

Fig 10.05

10.04 Histograms

Histograms in contrast to the bar chart enables more data to be shown. The width of each bar can be used to record the number of issues that have occurred.

Example 04

The same data as for fig 10.05 however the average temperature for each day can be recorded. The method is to add all of the temperatures per day for each month and then calculate the average.

The width of the bar is scaled to the number of days – fig 10.06.

January maximum temperatures

1st	$7°$	6th	$3°$	11th	$5°$	16th	$4°$	21st	$3°$	26th	$5°$
2nd	$6°$	7th	$2°$	12th	$6°$	17th	$3°$	22nd	$5°$	27th	$5°$
3rd	$5°$	8th	$3°$	13th	$7°$	18th	$2°$	23rd	$7°$	28th	$4°$
4th	$5°$	9th	$4°$	14th	$7°$	19th	$2°$	24th	$7°$	29th	$7°$
5th	$4°$	10th	$5°$	15th	$7°$	20th	$3°$	25th	$5°$	30th	$7°$
										31st	$6°$
27		17		32		14		27		34	

Add each of the sub-totals: 27 + 17 + 32 + 14 + 27 + 34 = 151°

Divide the total by the number of days to produce an average.

151 / 31 = 4.87° average per day. The histogram for January would be 31 units wide and 4.87 units high. The area would be 31 x 4.87 = 151 units. Note there are no set sizes for the units.

They can be any size that can be shown on the graph as long as the same sized units are used for each month. The size of the vertical units does not have to be the same scale as for the horizontal units.

If the data is to be shown on an A3 sheet landscape then the unit widths can be wider than if it is to be presented on A4 paper portrait.

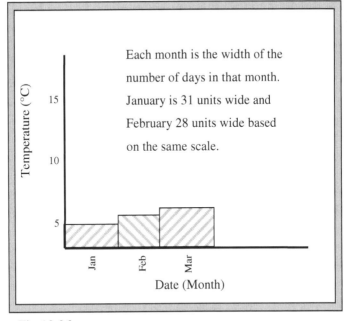

Fig 10.06

Note the graph now indicates the temperature for January as being cooler on a histogram than on a bar chart. The reason is the bar chart is showing the maximum temperature noted during the month and the histogram is showing the average temperature for that month.

Histogram blocks should be presented to a suitable scale and the areas can be easily compared. Where the data is constant such as every day throughout the year then the blocks should butt up against each other as shown in fig 10.06.

10.05 *Pie charts*

There are several different versions of the pie chart; as a single disk, isometric single disk, fragment pie and so on. The original concept is to present data as 100% being a pie. It is then divided up into percentages for each issue.

To hand draw the pie accurately the percentages for each issue need to be converted to an angle and a radius line either side.

Example 10.07

To present the following data as a pie chart the information must be converted into degrees.

The total is 600 and there are 410 light grey items, 145 dark grey items and 45 striped items. (410 + 145 + 45 = 600).

Stage 01

Convert each total to a percentage of the whole (600).

(410/600) x 100 = 68.33%

(145/600) x 100 = 24.17%

(45/600) x 100 = 7.5%

Check they total 100%. 68.33 + 24.17 + 7.5 = 100.

Stage 02

Convert the percentages into degrees.

(360 x 68.33) / 100 = 245.988°

(360 x 24.17) / 100 = 87.012°

(360 x 7.5) / 100 = 27°

Check the % add up to 360°

$$
\begin{array}{r}
245.988° \\
87.012° \\
\underline{27.000°} \\
360.00°
\end{array}
$$

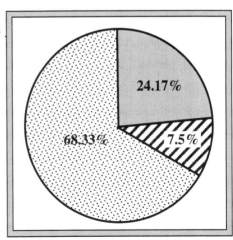

Fig 10.07

11.00 *Spreadsheets*

Perhaps the most useful tool for the construction industry. The principle is similar to the axes on a graph. The x axis runs from left to right (the columns) and identified by letters. The y axis runs north to south (the rows) identified by numbers.

To identify each cell a letter followed by a number is shown top left of the ribbon – fig 11.01.

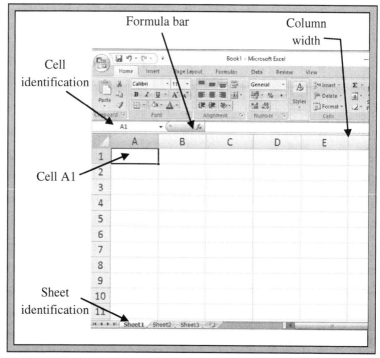

Fig 11.01

11.01 *Preparing a spreadsheet*

Open a new spreadsheet and save it either to a USB or the desktop. Save your work every few minutes in case either the computer locks up or you get called away from the machine. Losing your work is very frustrating and such a waste.

The file name will appear at the top of the screen.

Designing the data on the sheet can save time later. It is always easier to put the data in full and trim afterwards. The first example can be used to find the area and volume of a room.

Type 'Room area & volume' into cell A1 fig 11.02

As it is a title enlarge the font size to 14. Select cells A1, A2 and A3 and press the merge button. Shading and outlining the cells helps make the heading stand out. Why bother with that? Later when very large complex spreadsheets are being used it makes finding the data again easier.

Type in 'Length (m)' into cell A2, Width (m) in cell B2 and Area (m^2) into cell C2. To make the 2 smaller to show it is squared highlight the number only on the top ribbon and right click the mouse. The format cells menu should appear and select superscript and press ok. Those cells could be underlined to show they are headings.

Highlight cells A2, B2 and C2, right click and copy paste into cells A5 (the other cells will automatically fill). Drag the contents of C5 into D5 and highlight the superscript 2 in the top ribbon and type in 3 then enter. Into cell C5 type 'Height (m)'.

Writing formulae into cells requires an = sign first. That enables the cell to receive a formula.

Enter the following in cell C3: =A3*B3

Then into cell D6: =A6*B6*C6

Then save your work.

Into cell A6 enter: =A3

Into B6 enter: =B3

Highlight cells A3, B3 and shade, then shade C6. They will be the cells where data will be input. Try it out.

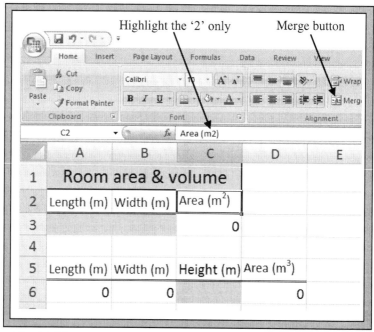

Fig 11.02

By linking specific cells the raw data only needs to be entered once in the shaded cells. Try changing the data in cell A3 and watch the results in both C3 (area) and D6 (volume).

Putting a complex formulae in one cell can be a source of error. An easier method is to use several cells with a small amount of data or formulae and check they work. Then combine the cell information into another cell if it all works.

If the cell has ###### it means the number is too big to show. Place the mouse cursor over the joint for column width (fig 11.01) and double click left mouse button quickly. The column width will automatically become wide enough to see the whole cell width.

11.02 *Formatting a spreadsheet*

Example 02

Formatting cells:

- Currency
- Dates
- Sum
- Sorting

Select a new spreadsheet using the tabs at the bottom of the screen. Double click the tab quickly and you will be able to type whatever identification you want. Try 'Example 2'.

Double click the tab for the first sheet used and type: 'Areas & Vol'. It makes it easier to find later.

Sheet 'Example 2'; Type in cell A1: Date, B1: Supplier, C1: Item, D1:Gross, E1: Nett, F1: VAT (20%)

Currency – any cell, row or column can become currency by selecting the cells, right click mouse and the format menu should drop down. Select currency and then 'ok'.

Highlight cells D2 and E2 down to D10 and E10 and select currency, 'ok'.

If you have invoices that include VAT it may be useful to automatically calculate the amount of VAT included in the invoice. A simple formula for cell F2: =(D2/120)*20

The math's: divide the gross that includes 20% VAT into 120 parts. That calculation is in brackets so it will be completed first. Then to multiply the result by 20 to show the amount of VAT.

Now type the formula to subtract the VAT from the gross figure into cell E2: =D2-F2

Dates can be formatted making it easier to present the data. Select the cell, row or column and right click the mouse. Select 'Format cells' from the drop down menu then 'date'.

Select the format you want and then press 'ok'. Using the numeric keyboard enter a date into a cell: 1/5/17 and it will convert to the chosen format.

Sum – automatic totalling can be put in a cell as a formula or use the Sigma button (Σ) in the 'editing' menu.

Formula method for cell D10. Type in: =sum(D2:D9)

Click on D10. Using the mouse place the cursor over the small square at the bottom right of the cell. The cursor should become a plus sign.

Now left click the mouse and drag the square to the right across E10 and F10 and release the button. E10 and F10 will have a similar formula in relation to each column.

Using the same method highlight cells E2 and F2 and drag the small square down to F9 and release. All the cells from E2 and F2 down to E9 and F9 will be populated.

Enter more data in the relevant cells of columns A, B, C and D only. The other data will automatically populate and all the calculations will be completed.

Sort – is a useful command that enables all or part of the data to be put into a required order.

In the example the dates are not in chronological order. To sort in descending date order use the mouse to highlight the rows to be sorted using the numbers on the left side of the screen.

Release the mouse button and the selected area should be shaded grey – fig 11.04.

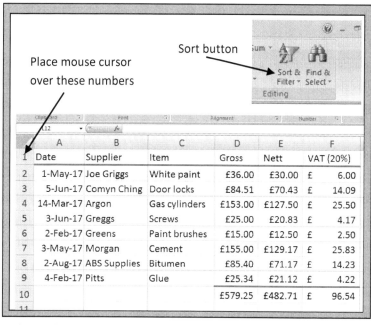

Fig 11.03

With only those rows highlighted click the left mouse button on 'Sort' and then 'custom sort' from the drop down menu.

The headings will be shown on a drop down menu under the heading 'Column'. Select 'date' and press 'ok'.

All of the data will be shuffled into the chronological order but the totals will have remained the same.

Using the same method it is possible to sort the data in date order ascending or descending, in order of suppliers or items in alphabetical order.

	A	B	C	D	E	F
	A2	▾	f_x 02-02-2017			
1	Date	Supplier	Item	Gross	Nett	VAT (20%)
2	2-Feb-17	Greens	Paint brushes	£15.00	£12.50	£ 2.50
3	4-Feb-17	Pitts	Glue	£25.34	£21.12	£ 4.22
4	14-Mar-17	Argon	Gas cylinders	£153.00	£127.50	£ 25.50
5	1-May-17	Joe Griggs	White paint	£36.00	£30.00	£ 6.00
6	3-May-17	Morgan	Cement	£155.00	£129.17	£ 25.83
7	3-Jun-17	Greggs	Screws	£25.00	£20.83	£ 4.17
8	5-Jun-17	Comyn Ching	Door locks	£84.51	£70.43	£ 14.09
9	2-Aug-17	ABS Supplies	Bitumen	£85.40	£71.17	£ 14.23
10				£579.25	£482.71	£ 96.54
11						

Fig 11.04

11.03 Angles and trigonometry

Spreadsheets are commonly used for recording and calculating surveying data. Total stations and some digital levels data can be recorded and downloaded in spreadsheet format.

Angles can be entered and used in calculations however they must be converted to radians.

11.04 *Degrees to Radians*

Starting a new spreadsheet label the bottom tab 'Angles'.

Type in cell A1: Degrees to radians

Change the font size to 14 and embolden the heading. Change the heading to cover 4 cells (A1 to D1) by highlighting them and press the 'merge' button – fig 11.02.

Type 'Angle (Degrees) in cell A2, and Angle (Radians) in cell B2. Change their font size to 8.

The text is still too big for the cells therefore select A2 and then click on 'wrap text' in the 'alignment group'. Repeat for cell B2.

Highlight both A2 and B2 and press the 'centring' button also in the alignment group – fig 11.05.

11.05 *RADIANS conversion formula*

The formula for conversion is already in the software. It only needs activating. Shade cell A3. Then in cell B3 type in the formula: =(RADIANS(A3))

Now try the sheet; type in 180 in cell A3 and the result will show in B3. It is 3.141593 otherwise known as π. That proves the cells are working correctly.

11.06 *Degrees to Sine radians*

Highlight cell A1 and copy paste into cell A5. Cells A5 to D5 will automatically merge. Click on the cell and then the top ribbon (formula ribbon) between the words 'to' and 'radians'. Enter the word 'Sine' and press enter.

Copy paste cells A2 and B2 into A6 and B6. Adjust the text in cell B6 to read Angle (Sine Radians).

The formula for cell B7 is: =SIN(RADIANS(A7)). Shade cell A7.

Enter an angle in degrees or decimal degrees and the sine of the angle will be shown in cell B7.

It is easy to check using the calculator. Press the 'sin' button and enter the same angle and close brackets and press =.

11.07 *Degrees to Cosine radians*

Prepare the cells as previously described and adjust to read Cosine Radians.

The formula for cell B11: =COS(RADIANS(A11))

Check with a calculator to ensure the answer is correct.

11.08 *Degrees to Tangent radians*

Prepare the cells as previously described and adjust to read Tan Radians. The formula for cell B11: =TAN(RADIANS(A14))

Check with a calculator to ensure the answer is correct.

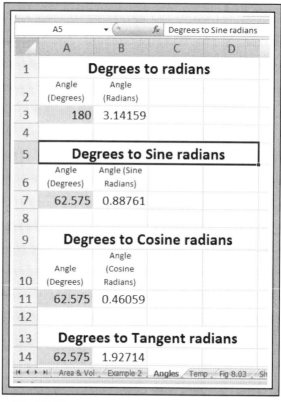

Fig 11.05

11.09 *Spreadsheet Trigonometry*

To carry out calculations using trigonometry enter the headings for the processes and enter the formulae.

Example 01

Based on the example used in chapter 8: fig 8.04.

Name the sheet tab 'Fig 8.04' and save the sheet. The formula used was Tan α = Opp / Adj. It was re-arranged (transposed) to: Adj = Opp / Tan α

For a spreadsheet it is easier just to arrange the cells to match the transposed formula so cell A1 type: Adj (m)

It is useful to put the units in brackets partly as a reminder and it helps anyone checking or using the sheet.

Cell B1: Height (m)

Cell C1: $\angle$ Degrees

Cell D1: Tan α

They are the headings. Change to font size 8 or 10, centre and underline. As an option shade cells B2 and C2 as they are the only cells that data will be entered.

The angle needs to be converted into radians therefore in cell D2 type: =TAN(RADIANS(C2))

Putting the cell reference (C2) in brackets will convert any angle entered in cell C2 to Tan radians.

The 'adjacent' is the unknown measurement therefore into cell A2 type: =B2/D2

The written transposed formula is: adj = opp / Tan α.

The spreadsheet version is:

Cell A2 is the 'adj' and the result of cell B2 'opp', (the height in metres) divided by D2 (tangent of the angle in radians).

Try changing the height to 3.0 and note the adjacent changes automatically.

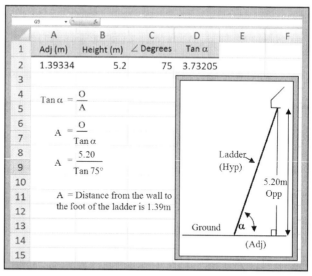

Fig 11.06

11.10 *Spreadsheet chart conversion*

Data on a spreadsheet can be converted into various charts and graphs. It is important to enter the data in a format that the software can use though.

Example 01

Temperature comparisons per day have been recorded.

	Sun	Mon	Tue	Wed	Thu	Fri	Sat
Temp (°C)	18	14	18	25	27	22	15

Name and save the spreadsheet.

In cell B1: Sun

Click the mouse on cell B1 then move the mouse icon over the small square at the bottom right hand corner and left click mouse and hold it down.

Now drag the small square to the right and release when over cell H1. The cells will be populated with the consecutive days of the week. Centre the cells using the button in alignment group.

An alternative method for centring is to right click the mouse on the cell to be formatted, choose 'format cells from the drop down menu. From that menu select the formatting required.

Cell A2: Temp (°C)

There are at least two ways to get the degrees sign; hold down the 'Alt' key and type 248 on the numeric keyboard and then release the 'Alt' key. Alternatively select 'Insert', 'Symbol' and then choose the degree symbol.

Add shading and lines, and then enter the data as shown above.

Highlight cells A1 and A2 through to H1 and H2 and click on Insert – Charts – Line – 2D. The chart appears on the spreadsheet. Using the corner handles it can be dragged to the required size.

A useful feature is the data is still live meaning it can be changed and the line graph changes accordingly – fig 11.07.

Using the same technique other charts can be produced.

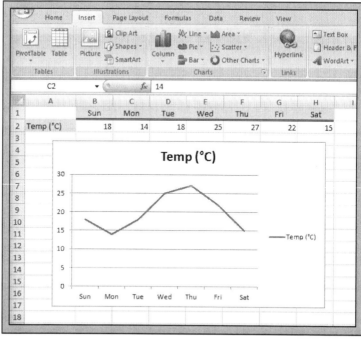

Fig 11.07

Index